Springer Proceedings in Mathematics & Statistics

Volume 57

For further volumes:
http://www.springer.com/series/10533

Springer Proceedings in Mathematics & Statistics

This book series features volumes composed of select contributions from workshops and conferences in all areas of current research in mathematics and statistics, including OR and optimization. In addition to an overall evaluation of the interest, scientific quality, and timeliness of each proposal at the hands of the publisher, individual contributions are all refereed to the high quality standards of leading journals in the field. Thus, this series provides the research community with well-edited, authoritative reports on developments in the most exciting areas of mathematical and statistical research today.

Clara Grácio • Daniele Fournier-Prunaret
Tetsushi Ueta • Yoshifumi Nishio
Editors

Nonlinear Maps and their Applications

Selected Contributions from the NOMA 2011 International Workshop

Springer

Editors
Clara Grácio
Colégio Luís António Verney
University of Évora CIMA-UE
Évora, Portugal

Tetsushi Ueta
Tokushima University
Tokushima, Japan

Daniele Fournier-Prunaret
LAAS-INSA
Toulouse, France

Yoshifumi Nishio
E. E. Eng.
Tokushima University
Tokushima, Japan

ISSN 2194-1009 ISSN 2194-1017 (electronic)
ISBN 978-1-4614-9160-6 ISBN 978-1-4614-9161-3 (eBook)
DOI 10.1007/978-1-4614-9161-3
Springer New York Heidelberg Dordrecht London

Library of Congress Control Number: 2013958084

Mathematics Subject Classification: 37-xx, 92Bxx, 37Fxx, 91-xx

Printed on acid-free paper

Springer is part of Springer Science+Business Media (www.springer.com)

Preface

The articles in this volume of the Springer Proceedings in Mathematics, titled *Nonlinear Maps and Their Applications*, has its origins in the Third International Workshop on Nonlinear Maps and Their Applications (NOMA 2011), which took place in Évora, Portugal, September 15–16, 2011. The workshop was organized and hosted by the University of Évora and the CIMA Research Center.

This conference follows in the sequence of Workshops on Nonlinear Maps and Their Applications, which began in Toulouse 2007 and continued in Urbino in 2009.

The objective of the NOMA 2011 conference was to provide a forum in the field of dynamical systems. It was open to theoretical studies as well as applications. The interaction and knowledge exchange among mathematicians, physicists, engineers, and other specialists as well as young researchers from the nonlinear sciences was very fruitful and gave rise to new developments in this area, in the pleasant atmosphere provided by the town of Évora.

The conference consisted of invited lectures given by Francisco Balibrea (University of Murcia, Spain), Ton Coolen (King's College London, UK), José Fernando Mendes (University of Aveiro, Portugal), Erik Mosekilde (Technical University of Denmark, Denmark), Iryna Sushko (National Academy of Sciences of Ukraine), Soumitro Banerjee (Indian Institute of Technology, Kharagpur, India), and Alberto Pinto (University of Porto, Portugal) as well as over 50 contributed lectures by representatives of 15 different countries (Portugal, Japan, France, UK, Italy, Spain, Ukraine, Germany, India, Russia, Norway, Denmark, Sweden, Poland, and Algeria).

This volume is a compilation of selected papers presented at NOMA that have been accepted for publication. The diversified nature of the presentations is reflected in the spectrum of the 16 papers published here.

The editors would like to thank all conference participants, organizers, and authors who submitted papers for this volume for their valuable contribution, as well as the reviewers for their expertise.

Évora, Portugal Clara Grácio
Toulouse, France Daniele Fournier-Prunaret
Tokushima, Japan Tetsushi Ueta
Yoshifumi Nishio

Contents

Chapter 1
Renormalization of Circle Diffeomorphism Sequences and Markov Sequences

João P. Almeida, Alberto A. Pinto, and David A. Rand

Abstract We show a one-to-one correspondence between circle diffeomorphism sequences that are C^{1+} n-periodic points of renormalization and smooth Markov sequences.

1.1 Introduction

Following [2–7, 15, 19–23], we present the concept of renormalization applied to circle diffeomorphism sequences. These concepts are essential for extending the results presented in [19, 20] to all Anosov diffeomorphisms on surfaces, i.e., for proving a one-to-one correspondence between C^{1+} conjugacy classes of Anosov diffeomorphisms and pairs of C^{1+} circle diffeomorphism sequences that are C^{1+} n-periodic points of renormalization (see also [1, 18–20]). The main point in this paper is to establish the existence of a one-to-one correspondence between C^{1+} circle diffeomorphism sequences that are C^{1+} n-periodic points of renormalization and smooth Markov sequences. This correspondence is a key step in passing from

J.P. Almeida (✉)
LIAAD-INESC TEC and Department of Mathematics, School of Technology and Management, Polytechnic Institute of Bragança, Campus de Santa Apolónia, Ap. 1134, 5301-857 Bragança, Portugal
e-mail: jpa@ipb.pt

A.A. Pinto
LIAAD-INESC TEC and Department of Mathematics, University of Porto, Rua do Campo Alegre, 4169-007 Porto, Portugal
e-mail: aapinto@fc.up.pt

D.A. Rand
Warwick Systems Biology and Mathematics Institute, University of Warwick, Coventry CV4 7AL, UK
e-mail: dar@maths.warwick.ac.uk

C. Grácio et al. (eds.), *Nonlinear Maps and their Applications*, Springer Proceedings in Mathematics & Statistics 57, DOI 10.1007/978-1-4614-9161-3_1,

circle diffeomorphisms to Anosov diffeomorphisms because the Markov sequences encode the smooth information of the expanding and contracting laminations of the Anosov diffeomorphisms [8–14, 16, 17].

1.2 Circle Diffeomorphisms

Let $\mathbf{a} = (a_i)_{i=0}^{\infty}$ be a sequence of positive integers and let $\gamma(\mathbf{a}) = 1/(a_0 + 1/(a_1 + 1/\cdots))$. For every $i \in \mathbb{N}_0$, let $\gamma_i = \gamma_i(\mathbf{a}) = 1/(a_i + 1/(a_{i+1} + 1/\cdots))$ and let $\mathbb{S}_i$ be a *counterclockwise oriented circle* homeomorphic to the circle $\underline{\mathbb{S}}_i = \mathbb{R}/(1+\gamma_i)\mathbb{Z}$.

An *arc* in $\mathbb{S}_i$ is the image of a nontrivial interval I in $\mathbb{R}$ by a homeomorphism $\alpha : I \to \mathbb{S}_i$. If I is closed (respectively open), we say that $\alpha(I)$ is a *closed* (respectively *open*) *arc* in $\mathbb{S}_i$. We denote by (a,b) (respectively $[a,b]$) the positively oriented open (respectively closed) arc on $\mathbb{S}_i$ beginning at the point $a \in \mathbb{S}_i$ and ending at the point $b \in \mathbb{S}_i$. A C^{1+} atlas $\mathscr{A}_i$ in $\mathbb{S}_i$ is a set of charts such that (1) every small arc of $\mathbb{S}_i$ is contained in the domain of some chart in $\mathscr{A}_i$, and (2) the overlap maps are $C^{1+\alpha}$ compatible, for some $\alpha > 0$.

Let $\underline{\mathscr{A}}_i$ denote the affine atlas whose charts are isometries with respect to the usual norm in $\underline{\mathbb{S}}_i$. Let the *rigid rotation* $\underline{g}_i : \underline{\mathbb{S}}_i \to \underline{\mathbb{S}}_i$ be the affine homeomorphism with respect to the atlas $\underline{\mathscr{A}}_i$ with rotation number $\gamma_i/(1+\gamma_i)$.

A homeomorphism $h : \underline{\mathbb{S}}_i \to \mathbb{S}_i$ is *quasisymmetric* if there exists a constant $C > 1$ such that for each two arcs I_1 and I_2 of $\underline{\mathbb{S}}_i$ with a common endpoint and such that $|I_1|_i = |I_2|_i$, we have $|h(I_1)|_i/|h(I_2)|_i < C$, where the lengths are measured in the charts of $\underline{\mathscr{A}}_i$ and $\mathscr{A}_i$.

A C^{1+} *circle diffeomorphism sequence* $(g_i, \mathbb{S}_i, \mathscr{A}_i)_{i=0}^{\infty}$ is a sequence of triples $(g_i, \mathbb{S}_i, \mathscr{A}_i)$ with the following properties: (1) $g_i : \mathbb{S}_i \to \mathbb{S}_i$ is a $C^{1+\alpha}$ diffeomorphism with respect to the $C^{1+\alpha}$ atlas $\mathscr{A}_i$ for some $\alpha > 0$; and (2) g_i is quasisymmetric conjugate to the rigid rotation $\underline{g}_i$ with respect to the atlas $\underline{\mathscr{A}}_i$.

We denote the C^{1+} circle diffeomorphism $(g_i, \mathbb{S}_i, \mathscr{A}_i)$ by g_i. In particular, we denote the rigid rotation $(\underline{g}_i, \underline{\mathbb{S}}_i, \underline{\mathscr{A}}_i)$ by $\underline{g}_i$.

1.2.1 Horocycles

Let us mark a point in $\mathbb{S}_i$, which we will denote by $0_i \in \mathbb{S}_i$. Let $S_i^0 = [0_i, g_i(0_i)]$ be the oriented closed arc in $\mathbb{S}_i$ with endpoints 0_i and $g_i(0_i)$. For every $k \in \{0, \ldots, a_i\}$, let $S_i^k = [g_i^k(0_i), g_i^{k+1}(0_i)]$ be the oriented closed arc in $\mathbb{S}_i$ with endpoints $g_i^k(0_i)$ and $g_i^{k+1}(0_i)$ and such that $S_i^k \cap S_i^{k-1} = \{g_i^k(0)\}$. Let $S_i^{a_i+1} = [g_i^{a_i+1}(0_i), 0_i]$ be the oriented closed arc in $\mathbb{S}_i$ with endpoints $g_i^{a_i+1}(0_i)$ and 0_i.

We introduce an *equivalence relation* $\sim$ in $\mathbb{S}_i$ by identifying the $a_i + 1$ points $g_i(0), \ldots, g_i^{a_i+1}(0)$ and form the topological space $H_i(\mathbb{S}_i, g_i) = \mathbb{S}_i/\sim$. We take the orientation in H_i as the reverse orientation of the one induced by $\mathbb{S}_i$. We call this oriented topological space the *horocycle*, and we denote it by $H_i = H_i(\mathbb{S}_i, g_i)$.

We consider the quotient topology in H_i. Let $\pi_{g_i} : \mathbb{S}_i \to H_i$ be the natural projection. The point

$$\xi_i = \pi_{g_i}(g_i(0_i)) = \cdots = \pi_{g_i}(g_i^{a_i+1}(0_i)) \in H_i$$

is called the *junction* of the horocycle H_i. For every $k \in \{0,\ldots,a_i\}$, let $S^k_{i,H} = S^k_{i,H}(\mathbb{S}_i, g_i) \subset H_i$ be the projection by π_{g_i} of the closed arc S^k_i. Let $R_i\mathbb{S}_i = S^0_{i,H} \cup S^{a+1}_{i,H}$ be the *renormalized circle* in H_i. The horocycle H_i is the union of the renormalized circle $R_i\mathbb{S}_i$ with the circles $S^k_{i,H}$ for every $k \in \{1,\ldots,a_i\}$.

A *parameterization* in H_i is the image of a nontrivial interval I in $\mathbb{R}$ by a homeomorphism $\alpha : I \to H_i$. If I is closed (respectively open), we say that $\alpha(I)$ is a *closed* (respectively *open*) *arc* in H_i. A *chart* in H_i is the inverse of a parameterization. A *topological atlas* $\mathscr{B}$ on the horocycle H_i is a set of charts $\{(j,J)\}$ on H_i with the property that every small arc is contained in the domain of a chart in $\mathscr{B}$, i.e., for every open arc K in H_i and $x \in K$, there exists a chart $\{(j,J)\} \in \mathscr{B}$ such that $J \cap K$ is a nontrivial open arc in H_i and $x \in J \cap K$. A C^{1+} *atlas* $\mathscr{B}$ *in* H_i is a topological atlas $\mathscr{B}$ such that the overlap maps are $C^{1+\alpha}$ and have $C^{1+\alpha}$ uniformly bounded norms, for some $\alpha > 0$.

Let $\mathscr{A}_i$ be a C^{1+} atlas on $\mathbb{S}_i$ in which $g_i : \mathbb{S}_i \to \mathbb{S}_i$ is a C^{1+} circle diffeomorphism. We are going to construct a C^{1+} atlas $\mathscr{A}_i^H$ on H_i that we call the *extended pushforward* $\mathscr{A}_i^H = (\pi_{g_i})_* \mathscr{A}_i$ of the atlas $\mathscr{A}_i$ on $\mathbb{S}_i$. If $x \in H_i \backslash \{\xi_i\}$, then there exists a sufficiently small open arc $J \subset H_i$ containing x such that $\pi_{g_i}^{-1}(J)$ is contained in the domain of some chart $(I, \hat{\imath})$ of $\mathscr{A}_i$. In this case, we define $(J, \hat{\imath} \circ \pi_{g_i}^{-1})$ as a chart in $\mathscr{A}_i^H$. If $x = \xi_i$ and J is a small arc containing ξ_i, then either (i) $\pi_{g_i}^{-1}(J)$ is an arc in $\mathbb{S}_i$ or (ii) $\pi_{g_i}^{-1}(J)$ is a disconnected set that is the union of two connected components.

In case (i), $\pi_{g_i}^{-1}(J)$ is connected, and it is contained in the domain of some chart $(I, \hat{\imath}) \in \mathscr{A}_i$. Therefore, we define $\left(J, \hat{\imath} \circ \pi_{g_i}^{-1}\right)$ as a chart in $\mathscr{A}_i^H$.

In case (ii), $\pi_{g_i}^{-1}(J)$ is a disconnected set that is the union of two connected arcs I^L_l and I^R_r of the form $I^L_l = (c^L_l, g^l_i(0)]$ and $I^R_r = [g^r_i(0), c^R_r)$, respectively, for all $l, r \in \{1,\ldots,a_i+1\}$. Let J^L_l and J^R_r be the arcs in H_i defined by $J^L_l = \pi_{g_i}(I^L_l)$ and $\pi_{g_i}(I^R_r)$ respectively. Then $J = J^L_l \cup J^R_r$ is an arc in H_i with the property that $J^L_l \cap J^R_r = \{\xi_i\}$, for every $l, r \in \{1,\ldots,a_i+1\}$. We call such an arc J an (l,r)-*arc*, and we denote it by $J_{l,r}$. Let $j_{l,r} : J_{l,r} \to \mathbb{R}$ be defined by

$$j_{l,r}(x) = \begin{cases} \hat{\imath} \circ \pi_{g_i}^{-1}(x) & \text{if } x \in J^R_r, \\ \hat{\imath} \circ g_i^{r-l} \circ \pi_{g_i}^{-1}(x) & \text{if } x \in J^L_l. \end{cases}$$

Let $(I, \hat{\imath}) \in \mathscr{A}_i$ be a chart such that $\pi_{g_i}(I) \supset J_{l,r}$. Then we define $(J_{l,r}, j_{l,r})$ as a chart in $\mathscr{A}_i^H$ (see Fig. 1.1). We call the atlas determined by these charts the *extended pushforward atlas of* $\mathscr{A}_i$, and by abuse of notation, we will denote it by $\mathscr{A}_i^H = (\pi_{g_i})_* \mathscr{A}_i$.

Let the marked point $\underline{0}_i$ in $\underline{\mathbb{S}}_i$ be the natural projection of $0 \in \mathbb{R}$ onto $\underline{0}_i \in \underline{\mathbb{S}}_i = \mathbb{R}/(1+\gamma_i)\mathbb{Z}$. Let $\underline{S}^0_i = [0_i, \underline{g}_i(0_i)]$ and $\underline{S}^k_i = [\underline{g}^k_i(0_i), \underline{g}^{k+1}_i(0_i)]$. Furthermore, let

$$\underline{H}_i = H_i(\underline{\mathbb{S}}_i, \underline{g}_i)\ ,\ \underline{S}^k_{i,H} = S^k_{i,H}(\underline{\mathbb{S}}_i, \underline{g}_i)\ ,\ R_i\underline{\mathbb{S}}_i = \underline{\mathbb{S}}^0_{i,H} \cup \underline{\mathbb{S}}^{a+1}_{i,H} \text{ and } \underline{\mathscr{A}}^{H_i} = \left(\pi_{\underline{g}_i}\right)_* \underline{\mathscr{A}}_i.$$

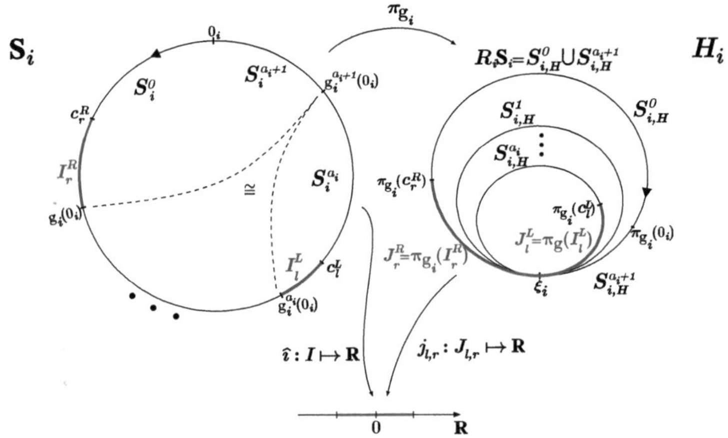

Fig. 1.1 The horocycle H_i and the chart $j_{l,r} : J_{l,r} \to \mathbb{R}$ in case (ii). The junction ξ_i of the horocycle is equal to $\xi_i = \pi_{g_i}(g_i(0_i)) = \cdots = \pi_{g_i}(g_i^{a_i}(0_i)) = \pi_{g_i}(g_i^{a_i+1}(0_i))$

1.3 Renormalization

The *renormalization of a C^{1+} circle diffeomorphism* g_i is the triple $(R_i g_i, R_i\mathbb{S}_i, R_i\mathscr{A}_i)$, where (1) $R_i\mathbb{S}_i$ is the renormalized circle with the orientation of the horocycle H_i, i.e., the reversed orientation of the orientation induced by $\mathbb{S}_i$; (2) the *renormalized atlas* $R_i\mathscr{A}_i = \mathscr{A}_i^H|_{R_i\mathbb{S}_i}$ is the set of all charts in $\mathscr{A}_i^H$ with domains contained in $R_i\mathbb{S}_i$; and (3) $R_i g_i : R_i\mathbb{S}_i \to R_i\mathbb{S}_i$ is the continuous map given by

$$R_i g_i(x) = \begin{cases} \pi_{g_i} \circ g_i^{a_i+1} \circ \left(\pi_{g_i}|_{S^0_{i,H}}\right)^{-1}(x) & \text{if } x \in S^0_{i,H}, \\ \pi_{g_i} \circ g_i \circ \left(\pi_{g_i}|_{S^{a_i+1}_{i,H}}\right)^{-1}(x) & \text{if } x \in S^{a_i+1}_{i,H}. \end{cases}$$

We denote the C^{1+} renormalization $(R_i g_i, R_i\mathbb{S}_i, R_i\mathscr{A}_i)$ of g_i by $R_i g_i$.

By construction, the renormalization $R_i \underline{g}_i$ of the rigid rotation $\underline{g}_i$ is affine conjugate to the rigid rotation $\underline{g}_{i+1}$. Hence from now on, we identify $(R_i\underline{g}_i, R_i\underline{\mathbb{S}}_i, R_i\underline{\mathscr{A}}_i)$ with $(\underline{g}_{i+1}, \underline{\mathbb{S}}_{i+1}, \underline{\mathscr{A}}_{i+1})$.

Recall that a C^{1+} *circle diffeomorphism* $g : \mathbb{S}_i \to \mathbb{S}_i$ is a $C^{1+\alpha}$ diffeomorphism with respect to a $C^{1+\alpha}$ atlas $\mathscr{A}$ on $\mathbb{S}_i$, for some $\alpha > 0$, that is quasisymmetric conjugate to a rigid rotation $\underline{g} : \mathbb{S}_i \to \mathbb{S}_i$ with respect to an affine atlas $\underline{\mathscr{A}}$ on $\underline{\mathbb{S}}_i$.

The renormalization $R_i g_i$ is a C^{1+} circle diffeomorphism quasisymmetric conjugate to the rigid rotation $\underline{g}_{i+1}$. Hence $R_i g_i$ is quasisymmetric conjugate to the C^{1+} circle diffeomorphism g_{i+1}. The marked point $0_i \in \mathbb{S}_i$ determines the marked point $0_{R_i\mathbb{S}_i} = \pi_{g_i}(0_i)$ in the circle $R_i\mathbb{S}_i$. Thus, there is a unique topological conjugacy h_i between $R_i g_i$ and g_{i+1} such that $h_i(0_{R_i\mathbb{S}_i}) = 0_{i+1}$ (see Fig. 1.2).

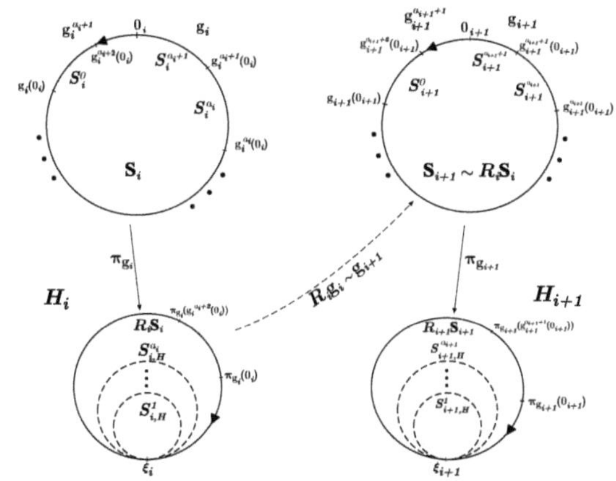

Fig. 1.2 The horocycles H_i and H_{i+1}, and the renormalized map $R_i g_i : R_i\mathbb{S}_i \to R_i\mathbb{S}_i$. Here $\xi_i = g_i(0_i) = \ldots = g_i^{a_i+1}(0_i)$, $\xi_{i+1} = g_{i+1}(0_{i+1}) = \cdots = g_{i+1}^{a_{i+1}+1}(0_{i+1}))$, and the map $R_i g_i$ is identified with g_{i+1}

A C^{1+} circle diffeomorphism g_0 determines a unique C^{1+} *renormalization circle diffeomorphism sequence* $\mathbf{R}(g_0) = (g_i, \mathbb{S}_i, \mathscr{A}_i)_{i=0}^{\infty}$ given by

$$(g_i, \mathbb{S}_i, \mathscr{A}_i) = (R_i \circ \cdots \circ R_0 g_0, R_i \circ \cdots \circ R_0 \mathbb{S}_0, R_i \circ \cdots \circ R_0 \mathscr{A}_0).$$

We note that the C^{1+} renormalization circle diffeomorphism sequence $\mathbf{R}(g_0)$ is a C^{1+} circle diffeomorphism sequence.

We say that a sequence $\mathbf{a} = (a_i)_{i=0}^{\infty}$ is *n-periodic* if n is the least integer such that $a_{i+n} = a_i$ for every $i \in \mathbb{N}_0$. We observe that given a n-periodic sequence of positive integers $\mathbf{a} = (a_i)_{i=0}^{\infty}$, $\gamma_i = \gamma_i(\mathbf{a}) = 1/(a_i + 1/(a_{i+1} + 1/\cdots))$ is equal to γ_{i+n} for every $i \in \mathbb{N}_0$. Hence there exists a topological conjugacy $\phi_i : \mathbb{S}_i \to \mathbb{S}_{i+n}$ such that

$$\phi_i \circ g_i = g_{i+n} \circ \phi_i,$$

because g_i and $g_{i+n} = R_{i+n} \circ \cdots \circ R_i g_i$ are C^{1+} circle diffeomorphisms with the same rotation number $\gamma_i = \gamma_{i+n}$.

We say that a sequence $\mathbf{R}(g_0)$ is a C^{1+} *n-periodic point of renormalization* if ϕ_i is C^{1+} for every $i \in \mathbb{N}$.

1.4 Markov Maps

Let $\mathbf{R}(g_0)$ be the renormalization circle diffeomorphism sequence associated to the C^{1+} circle diffeomorphism g_0. The *Markov map* $M_i : H_i \to H_{i+1}$ is given by

$$M_i(x) = \begin{cases} \pi_{g_{i+1}}(x) & \text{if } x \in R_i\mathbb{S}_i, \\ \pi_{g_{i+1}} \circ \pi_{g_i} \circ g_i^{-k} \circ \pi_{g_i}^{-1}(x) & \text{if } x \in S^k_{i,H_i}, \text{ for } k = 1, \ldots, a_i. \end{cases}$$

The *Markov sequence* $(M_i)_{i=0}^{\infty}(g_0)$ associated to a C^{1+} circle diffeomorphism g_0 is the sequence of Markov maps $M_i : H_i \to H_{i+1}$ for $i \in \mathbb{N}_0$. Two Markov sequences $(M_i)_{i=0}^{\infty}(g_0)$ and $(N_i)_{i=0}^{\infty}(g_0)$ are *quasisymmetric conjugate* if there is a sequence $(h_i)_{i=0}^{\infty}$ of quasisymmetric maps h_i such that $M_{i+1} \circ h_i = h_{i+1} \circ M_i$ for each $i \in \mathbb{N}_0$.

The *rigid Markov sequence* $(\underline{M}_i)_{i=0}^{\infty} = (M_i)_{i=0}^{\infty}(\underline{g}_0)$ is the Markov sequence associated to the rigid rotation $\underline{g}_0$. The rigid Markov maps $\underline{M}_i : \underline{H}_i \to \underline{H}_{i+1}$ are affine with respect to the atlases $\underline{\mathscr{A}}_i^H$ and $\underline{\mathscr{A}}_{i+1}^H$ (see Fig. 1.3).

The Markov sequence $(M_i)_{i=0}^{\infty}(g_0)$ has the following properties: (i) the Markov maps M_i are local $C^{1+\alpha}$ diffeomorphisms for some $\alpha > 0$, and (ii) the Markov sequence $(M_i)_{i=0}^{\infty}(g_0)$ is quasisymmetric conjugate to the rigid Markov sequence $(\underline{M}_i)_{i=0}^{\infty}(\underline{g}_0)$ because g_i is quasisymmetric conjugate to $\underline{g}_i$.

The *n-extended Markov sequence* $(\mathbf{M}_i)_{i=0}^{n-1}(g_0)$ is the sequence of the *n-extended Markov maps* $\mathbf{M}_i(g_0) : H_i \to H_i$ defined by

$$\mathbf{M}_i(g_0) = \phi_i^{-1} \circ M_{i+n} \circ \cdots \circ M_i.$$

We observe that a sequence $\mathbf{R}(g_0)$ is a C^{1+} *n-periodic point of renormalization* if and only if the n-extended Markov maps $\mathbf{M}_i : H_i \to H_i$ are C^{1+} for every $i \in \mathbb{N}$.

The *rigid n-extended Markov sequence* $(\underline{\mathbf{M}}_i)_{i=0}^{n-1} = (\mathbf{M}_i)_{i=0}^{n-1}(\underline{g}_0)$ is the n-extended Markov sequence associated to the rigid rotation $\underline{g}_0$. The rigid n-extended Markov maps $\mathbf{M}_0, \ldots, \mathbf{M}_{n-1}$ are affine with respect to the atlases $\underline{\mathscr{A}}_0^H, \ldots, \underline{\mathscr{A}}_{n-1}^H$, respectively, because the conjugacy maps $\underline{\phi}_i : \underline{\mathbb{S}}_i \to \underline{\mathbb{S}}_{i+n}$ are affine.

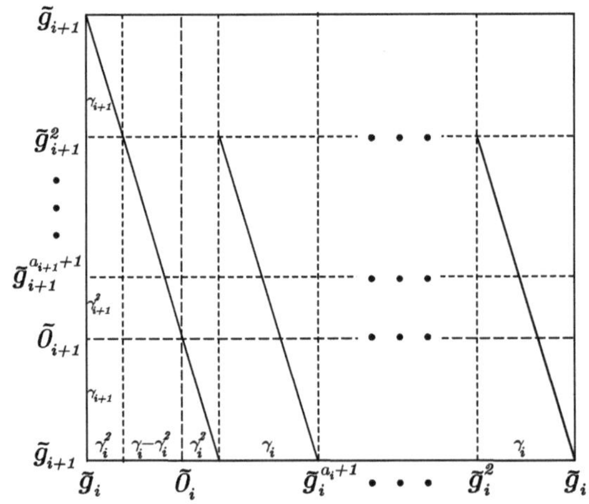

Fig. 1.3 A representation of the rigid Markov map $\underline{M}_i : \underline{H}_i \to \underline{H}_{i+1}$ with respect to the atlases $\underline{\mathscr{A}}_i^H$ and $\underline{\mathscr{A}}_{i+1}^H$, respectively. Here we represent by $\tilde{0}_i$ and $\tilde{0}_{i+1}$ the points $\pi_{\underline{g}_i}(\underline{0}_i)$ and $\pi_{\underline{g}_{i+1}}(\underline{0}_{i+1})$, respectively, and by $\tilde{g}_k^l$ the points $\pi_{\underline{g}_k} \circ \underline{g}_k^l(\underline{0}_k)$, for $k \in \{i, i+1\}$ and $l \in \{1, \ldots, a_k + 1\}$

If $\phi_i : \mathbb{S}_i \to \mathbb{S}_{i+n}$ is $C^{1+\alpha}$, then the n-extended Markov sequence $(\mathbf{M}_i)_{i=0}^{n-1}(g_0)$ has the following properties: (1) the n-extended Markov maps $\mathbf{M}_i$ are local $C^{1+\alpha}$ diffeomorphisms, for some $\alpha > 0$, because the Markov maps $M_0, \ldots M_{n-1}$ of the sequence $(M_i)_{i=0}^{n-1}(g_0)$ are local $C^{1+\alpha}$ diffeomorphisms, and (2) the n-extended Markov maps $\mathbf{M}_i$ are quasisymmetric conjugate to the rigid n-extended Markov maps $\underline{\mathbf{M}}_i$ because the Markov maps $M_0, \ldots M_{n-1}$ are quasisymmetric conjugate to $\underline{M}_0, \ldots, \underline{M}_{n-1}$.

Acknowledgements We acknowledge the financial support of LIAAD-INESC TEC through 'Strategic Project - LA 14 - 2013–2014' with reference PEst-C/EEI/LA0014/2013, USP-UP project, IJUP, Faculty of Sciences, University of Porto, Calouste Gulbenkian Foundation, FEDER, POCI 2010 and COMPETE Programmes and Fundação para a Ciência e a Tecnologia (FCT) through Project 'Dynamics and Applications', with reference PTDC/MAT/121107/2010. J. P. Almeida acknowledges the support from FCT, given through grant SFRH/PROTEC/49754/2009. Part of this research was done during visits by the authors to IMPA (Brazil), University of Säo Paulo (Brazil), University of Warwick (United Kingdom), Institut Henry Poincaré (France) and SUNY (USA). The authors thank them for their hospitality.

References

1. Almeida, J.P., Fisher, A.M., Pinto, A.A., Rand, D.A.: Anosov and circle diffeomorphisms. In: Peixoto, M., Pinto, A., Rand, D. (eds.) Dynamics Games and Science I. Springer Proceedings in Mathematics, pp. 11–23. Springer, Berlin (2011)
2. Arnol'd, V.I.: Small denominators I: on the mapping of a circle into itself. Investijia Akad. Nauk. Math. (Transl. Am. Math .Soc. 2nd Ser. **46**, 213–284) **25**(1), 21–96 (1961)
3. Coullet, P., Tresser, C.: Itération d'endomorphismes et groupe de renormalisation. J. de Phys. Colloque **539**, C5–25 (1978)
4. Herman, M.R.: Sur la conjugaison différentiable des difféomorphismes du cercle à des rotations. Publ. IHES **49**, 5–233 (1979)
5. Lanford, O.: Renormalization group methods for critical circle mappings with general rotation number. VIIIth International Congress on Mathematical Physics, pp. 532–536. World Science Publishing, Singapore (1987)
6. Masur, H.: Interval exchange transformations and measured foliations. Ann. Math. 2nd Ser. **115**(1), 169–200 (1982)
7. Penner, R.C., Harer, J.L.: Combinatorics of Train-Tracks. Princeton University Press, Princeton (1992)
8. Pinto, A.A., Rand, D.A.: Characterising rigidity and flexibility of pseudo-Anosov and other transversally laminated dynamical systems on surfaces. Warwick Preprint (1995)
9. Pinto, A.A., Rand, D.A.: Existence, uniqueness and ratio decomposition for Gibbs states via duality. Ergod. Theory Dynam. Syst. **21**, 533–543 (2001)
10. Pinto, A.A., Rand, D.A.: Smoothness of holonomies for codimension 1 hyperbolic dynamics. Bull. London Math. Soc. **34**, 341–352 (2002)
11. Pinto, A.A., Rand, D.A.: Teichmüller spaces and HR structures for hyperbolic surface dynamics. Ergod. Theory Dynam. Syst. **22**, 1905–1931 (2002)
12. Pinto, A.A., Rand, D.A.: Rigidity of hyperbolic sets on surfaces. J. London Math. Soc. **71**(2), 481–502 (2004)
13. Pinto, A.A., Rand, D.A.: Solenoid functions for hyperbolic sets on surfaces. In: Boris Hasselblat (ed.) Dynamics, Ergodic Theory and Geometry. Mathematical Sciences Research Institute Publications, vol. 54, pp. 145–178. Cambridge University Press, Cambridge (2007)
14. Pinto, A.A., Rand, D.A.: Train-tracks with C^{1+} self-renormalisable structures. J. Differ. Equat. Appl. **16**(8), 945–962 (2010)
15. Pinto, A.A., Sullivan, D.: The circle and the solenoid. Dedicated to Anatole Katok on the Occasion of his 60th Birthday. Discrete Contin. Dynam. Syst. A **16**(2), 463–504 (2006)

16. Pinto, A.A., Rand, D.A., Ferreira, F.: Cantor exchange systems and renormalization. J. Differ. Equat. **243**, 593–616 (2007)
17. Pinto, A.A., Rand, D.A., Ferreira, F.: Arc exchange systems and renormalization. J. Differ. Equat. Appl. **16**(4), 347–371 (2010)
18. Pinto, A.A., Rand, D.A., Ferreira, F.: Fine Structures of Hyperbolic Diffeomorphisms. Springer Monographs in Mathematics. Springer, New York (2010)
19. Pinto, A.A., Almeida, J.P., Rand, D.A.: Anosov and renormalized circle diffeomorphisms, pp. 1–33 (2012, submitted).
20. Pinto, A.A., Almeida, J.P., Portela, A.: Golden tilings. Trans. Am. Math. Soc. **364**, 2261–2280 (2012)
21. Rand, D.A.: Existence, nonexistence and universal breakdown of dissipative golden invariant tori. I. Golden critical circle maps. Nonlinearity **5**(3), 639–662 (1992)
22. Veech, W.: Gauss measures for transformations on the space of interval exchange maps. Ann. Math. 2nd Ser. **115**(2), 201–242 (1982)
23. Yoccoz, J.C.: Conjugaison différentiable des difféomorphismes du cercle dont le nombre de rotation vérifie une condition diophantienne. Ann. Scient. Éc. Norm. Sup. 4 série t **17**, 333–359 (1984)

Chapter 2
Examples of Lyapunov Exponents in Two-Dimensional Systems

Francisco Balibrea and M. Victoria Caballero

Abstract In this paper, we construct two examples of two-dimensional dynamical systems. In both examples, we have computed the Lyapunov exponents for the forward trajectory of $(0,0)$. In the first, such a forward trajectory has positive Lyapunov exponent but does not have sensitive dependence on initial conditions, while the second example has negative Lyapunov exponent but sensitive dependence on initial conditions.

2.1 Introduction

In recent years, Lyapunov exponents have been a powerful tool for understanding chaotic behavior and predictability in nonlinear discrete dynamical systems and time series obtained from models. But this tool introduces a host of difficulties of interpretation in the sense of deciding whether chaos and random effects are present or whether it is possible to determine the existence of external influences.

At the same time, it is a widespread practice, especially in experimental and applied dynamics, to associate the presence of a positive Lyapunov exponent with instability and a negative Lyapunov exponent with stability of orbits of a dynamical system. Stability and instability of orbits are defined in topological terms, while a Lyapunov exponent is a numerical characteristic calculated throughout the orbit. However, these interpretations have no firm mathematical foundation unless some restrictions on the maps describing the system are introduced. To illustrate this, in [3, 4], two interval maps have been constructed: one has an orbit with positive Lyapunov exponent but is Lyapunov stable, while the otherhas an orbit with negative

F. Balibrea (✉) • M.V. Caballero
Universidad de Murcia, Campus de Espinardo, 30100 Murcia, Spain
e-mail: balibrea@um.es; mvictori@um.es

C. Grácio et al. (eds.), *Nonlinear Maps and their Applications*, Springer Proceedings in Mathematics & Statistics 57, DOI 10.1007/978-1-4614-9161-3_2,

Lyapunov exponent and sensitive dependence on initial conditions (s.d.i.c.). Such examples have been improved in some sense in [1].

Using the examples from [1], in this paper we construct examples of two discrete two-dimensional dynamical systems with similar behaviors to the former. Before constructing them, it is necessary to begin with some introductory definitions.

2.2 Definitions and Preliminary Results

Given a dynamical systems (X, f), where X is a compact space and f a continuous map from X into itself, the trajectory of a point $x \in X$ is the sequence $(f^n(x))_{n=0}^{\infty} = (x_n)_{n=0}^{\infty}$, where as usual, f^n denotes the n-iteration of f acting on x. Two trajectories starting from nearby initial conditions can diverge or converge in certain directions under f as n grows. Lyapunov exponents provide a means of quantifying the expansion or contraction of such trajectories in different directions.

Let $X \subset \mathbb{R}^m$ and let d be any metric on it. If $(x_n)_{n=0}^{\infty}$ and $(x'_n)_{n=0}^{\infty}$ are two trajectories starting from nearby initial states x_0 and x'_0, we write $\delta x_n = x'_n - x_n$. If f has continuous partial derivatives in every x_i, then iterating the map, we have a linear approximation in which $Df(x)$ denotes the differential of the map f at the point x with the (i, j) element of this matrix given by $\frac{\partial f_i}{\partial x_j}$, where f_i and x_j are the components of f and x in local coordinates on X, and

$$Df^n(x_0) = Df(x_{n-1})Df(x_{n-2}) \cdots Df(x_1)Df(x_0),$$

where δx_n represents the separation of these orbits after n iterations of the map f:

$$\delta x_n \simeq Df^n(x_0)\delta x_0 = \prod_{i=0}^{n-1} Df(x_i)\delta x_0.$$

The matrix $(Df^n(x_0)^t)(Df^n(x_0))$ has m eigenvalues given by $\mu_i(n, x_0)$, where $i = 1, 2, \ldots, m$, such that $\mu_1(n, x_0) \geq \mu_2(n, x_0) \geq \cdots \geq \mu_m(n, x_0)$.

Definition 1. The ith local Lyapunov exponent at x_0 is defined by

$$\lambda_i(x_0) = \lim_{n \to \infty} \frac{1}{2n} \log(|\mu_i(n, x_0)|)$$

if this limit exists.

In [5] are stated conditions for the existence of such a limit. Now we recall the notions of instability and stability in the Lyapunov sense.

Definition 2 (Sensitive dependence or Lyapunov instability). The forward orbit $(x_n)_{n=0}^{\infty}$ has sensitive dependence on initial conditions (s.d.i.c.) if there exists $\varepsilon > 0$ such that for every $\delta > 0$, there exists y with $d(x_0, y) < \delta$ and $N \geq 0$ such that $d(f^N(y), f^N(x_0)) \geq \varepsilon$.

Definition 3. The forward orbit $(x_n)_{n=0}^{\infty}$ does not have s.d.i.c. or it is Lyapunov stable if for every $\varepsilon > 0$, there is $\delta > 0$ such that whenever $d(x_0, y) < \delta$, then $d(f^n(y), f^n(x_0)) < \varepsilon$ for all $n \geq 0$.

2.3 Two Two-Dimensional Examples

We propose two dynamical systems, one defined in $[0,1]^2$ that has a forward trajectory with a positive Lyapunov exponent but not having s.d.i.c., and other defined in $[0,1)^2$ that has a forward trajectory with a negative Lyapunov exponent but having s.d.i.c. The examples are two-dimensional versions of those mentioned in the introduction. The maps we are using are examples of permutation maps introduced in [2].

Example 1. We are going to obtain a continuous function $F = (f, g)$ in $[0,1]^2$ such that the forward trajectory of $(0,0)$ has a positive Lyapunov exponent but does not have s.d.i.c.

(a) The first component of F is the map $f : [0,1] \to [0,1]$ that was introduced in [3]:

$$f(x) = \begin{cases} 2x - 1 + \dfrac{1}{2^{n+1}}, & a_n < x \leq b_n, \, x \in [0,1], \\ \dfrac{5^{n+2} - 22}{2 \cdot 5^{n+2} - 11}(x - b_n) + 1 + \dfrac{2}{10^{n+1}} - \dfrac{1}{2^{n+1}}, & b_n < x \leq a_{n+1}, \\ 1, & x = 1, \end{cases}$$

with $a_n = 1 - 2^{-n} - 10^{-n-1}$, $b_n = 1 - 2^{-n} + 10^{-n-1}$, $n = 0, 1, 2, \ldots$.

(b) The second component of F is $g : [0,1] \to [0,1]$ such that

$$g(x) = \begin{cases} 3x + \dfrac{1}{2}, & 0 \leq x \leq \dfrac{1}{15}, \\ \dfrac{6}{127}x + \dfrac{7}{10} - \dfrac{2}{635}, & \dfrac{1}{15} < x \leq \dfrac{1}{2} - \dfrac{1}{100}, \\ 3x + \dfrac{1}{2} - \dfrac{5}{2^{n+1}}(2^n - 1), & a_n < x \leq b_n, \\ \dfrac{5^{n+2} - 33}{2 \cdot 5^{n+2} - 11}(x - b_n) + 1 + \dfrac{3}{10^{n+1}} - \dfrac{1}{2^{n+1}}, & b_n < x \leq a_{n+1}, \\ 1, & x = 1, \end{cases}$$

with $a_n = 1 - 2^{-n} - 10^{-n-1}$, $b_n = 1 - 2^{-n} + 10^{-n-1}$, $n = 1, 2, \ldots$.

The map $F(x,y)=(f(y),g(x))$ is continuous in $[0,1]^2$ because f and g are continuous in $[0,1]$. We consider the trajectory of $(0,0)$:

$$\{(0,0),(x_1,y_1),(x_2,y_2),\cdots\}=\left\{(1-\frac{1}{2^k},1-\frac{1}{2^k})\right\}_{k=0}^{\infty}$$

The map is differentiable at every point of this trajectory (except $(0,0)$). Since f and g are differentiable maps to the right of 0, we define

$$DF(0^+,0^+)=\begin{pmatrix}0 & \lim_{y\to 0^+} f(y)\\ \lim_{x\to 0^+} g(x) & 0\end{pmatrix}=\begin{pmatrix}0 & 2\\ 3 & 0\end{pmatrix}$$

and

$$DF(x_1,y_1)=DF^2(0,0)=\begin{pmatrix}6 & 0\\ 0 & 6\end{pmatrix},$$

$$DF(x_2,y_2)=DF^3(0,0)=\begin{pmatrix}0 & 12\\ 18 & 0\end{pmatrix},$$

that is, in general,

$$DF^{2n}(0,0)=\begin{pmatrix}6^n & 0\\ 0 & 6^n\end{pmatrix} \qquad DF^{2n-1}(0,0)=\begin{pmatrix}0 & 2\cdot 6^{n-1}\\ 3\cdot 6^{n-1} & 0\end{pmatrix}$$

for $n=1,2,\ldots$.

Now we compute the eigenvalues $(DF^k)^t(DF^k)$, where $k=2n$ or $k=2n-1$ with $n=1,2,\ldots$. Then we have, when $k=2n-1$,

$$(DF^{2n-1}(0,0))^t DF^{2n-1}(0,0)=\begin{pmatrix}3^2\cdot 6^{2(n-1)} & 0\\ 0 & 2^2\cdot 6^{2(n-1)}\end{pmatrix},$$

and the largest eigenvalue of such a matrix is $\mu_1(2n-1,(0,0))=3^2 6^n$, and when $k=2n$,

$$(DF^{2n}(0,0))^2=\begin{pmatrix}6^{2n} & 0\\ 0 & 6^{2n}\end{pmatrix},$$

whose eigenvalue is $\mu_1(2n,(0,0))=6^{2n}$.

Therefore, the Lyapunov exponent of the forward orbit with initial condition $(0,0)$ is

$$\lambda_1(0,0)=\lim_{k\to\infty}\frac{1}{2k}\log(|\mu_1(k,(0,0))|)=\log 6>0.$$

It is left to prove that the forward trajectory of $(0,0)$ does not have s.d.i.c.

To see this, let $\varepsilon>0$. We can chose $k>0$ such that $1/2^k<\varepsilon$. Since f' and g' are positive on the components of the forward orbit of $(0,0)$, with the maximum distance denoted by $|\cdot|$, we have that

$$|F^k(x,y) - F^k(0,0)| \leq |(\frac{1}{2^k}, \frac{1}{2^k})| = \frac{1}{2^k} < \varepsilon$$

for $n \geq k$ and $0 < |(x,y)| < \bar{\delta}$. It remains to prove that the last inequality holds for $n < k$. This is made by the continuity of F^j, since given $\varepsilon > 0$ there exists δ_j such that if $0 < |(x,y)| < \delta_j$, then $|F^j(x,y) - F^j(0,0)| < \varepsilon$ for $j = 1,\ldots,n-1$. Then if we take

$$\delta = \min\{\delta_1,\ldots,\delta_{n-1},\bar{\delta}\} \text{ and } 0 < |(x,y)| < \delta,$$

it follows that

$$|F^k(x,y) - F^k(0,0)| < \varepsilon$$

for all $k > 0$.

Example 2. We are going to obtain a continuous function $G = (f^2, g)$ in $[0,1)^2$ such that the forward trajectory of $(0,0)$ has a negative Lyapunov exponent but it does not have s.d.i.c.

(a) The first component of G is f^2, where the map $f : [0,1) \to [0,1)$ is defined in [3] by

$$f(x) = \begin{cases} \frac{1}{2}x + \frac{1}{2} & 0 \leq x < 7/16 \text{ or } a_n \leq x < b_n \\ (2^{n+1} - 4^{n+1} - 2^{-1})(x + 2^{-n} - 2 \cdot 4^{-n-1} - 1) & b_n \leq x < c_n \\ \frac{1 - 2^{-n-2} - 2 \cdot 4^{-n-3}}{2^{-n-1} - 9 \cdot 4^{-n-2}}(x + 2^{-n} - 2 \cdot 4^{-n-1} - 1) & c_n \leq x < a_{n+1} \end{cases}$$

where $a_n = 1 - 2^{-n} - 4^{-n-1}$, $b_n = 1 - 2^{-n} + 4^{-n-1}$, $c_n = 1 - 2^{-n} + 2 \cdot 4^{-n-1}$ for $n = 1,2,\ldots$.

(b) The second component is $g : [0,1) \to [0,1)$, defined by

$$g(x) = \begin{cases} 3x + \frac{1}{2} & 0 \leq x \leq \frac{1}{15} \\ \frac{6}{127}x + \frac{7}{10} - \frac{2}{635} & \frac{1}{15} < y \leq \frac{1}{2} - \frac{1}{100} \\ 3x + \frac{1}{2} - \frac{5}{2^{n+1}}(2^n - 1) & a_n < x \leq b_n \\ \frac{5^{n+2} - 33}{2 \cdot 5^{n+2} - 11}(x - b_n) + 1 + \frac{3}{10^{n+1}} - \frac{1}{2^{n+1}} & b_n < x \leq a_{n+1} \end{cases}$$

where $a_n = 1 - 2^{-n} - 10^{-n-1}$, $b_n = 1 - 2^{-n} + 10^{-n-1}$, for $n = 1,2,\ldots$.

The map $G(x,y) = (f^2(y), g(x))$ is continuous in $[0,1)^2$, since f and g are continuous in $[0,1)$. Let us consider the trajectory of $(0,0)$, denoted by

$$\{(0,0),(x_1,y_1),(x_2,y_2),\ldots\},$$

where

$$(x_{2n}, y_{2n}) = \left(1 - \frac{1}{2^{3n}}, 1 - \frac{1}{2^{3n}}\right) \quad \text{and } (x_{2n-1}, y_{2n-1}) = \left(1 - \frac{1}{2^{3n-1}}, 1 - \frac{1}{2^{3n-2}}\right)$$

for $n = 1, 2, \ldots$.

Similarly to the previous example, we have

$$DG(0,0) = \begin{pmatrix} 0 & 1/4 \\ 3 & 0 \end{pmatrix}, \qquad DG(x_1, y_1) = DG^2(0,0) = \begin{pmatrix} 3/4 & 0 \\ 0 & 3/4 \end{pmatrix},$$

$$DG(x_2, y_2) = DG^3(0,0) = \begin{pmatrix} 0 & 3/42 \\ 32/4 & 0 \end{pmatrix},$$

and in general we have

$$DG^{2n}(0,0) = \begin{pmatrix} (3/4)^n & 0 \\ 0 & (3/4)^n \end{pmatrix}, \qquad DG^{2n-1}(0,0) = \begin{pmatrix} 0 & \frac{3^{n-1}}{4^n} \\ \frac{3^n}{4^{n-1}} & 0 \end{pmatrix},$$

for $n = 1, 2, \ldots$.

Now we compute the eigenvalues of $(DG^k)^t \cdot DG^k$ when $k = 2n$ or $k = 2n-1$ with $n = 1, 2 \ldots$ Then when $k = 2n-1$, we have

$$(DG^{2n-1}(0,0))^t DG^{2n-1}(0,0) = \begin{pmatrix} \frac{3^{2n}}{4^{2(n-1)}} & 0 \\ 0 & \frac{3^{2(n-1)}}{4^{2n}} \end{pmatrix}$$

and $\mu_1(2n-1, (0,0)) = \dfrac{3^{2n}}{4^{2(n-1)}}$.

And when $k = 2n$, we have that $\mu_1(2n, (0,0)) = \dfrac{3^{2n}}{4^{2n}}$.

Then the Lyapunov exponent of the orbit through $(0,0)$ is

$$\lambda_1(0,0) = \lim_{k \to \infty} \frac{1}{2k} \log(|\mu_1(k, (0,0))|) = \log\left(\frac{3}{4}\right) < 0.$$

It is left to prove that the forward trajectory of $(0,0)$ has s.d.i.c. We use the maximum distance, which is represented by $|\cdot|$.

To do it, fix $\varepsilon = 3/8$ and take $k \geq 2$ such that $f^k < 1/2$ and $f^k(0) < 7/8$. Since for every $\delta > 0$ there exists (x,y) with $|(x,y)| < \delta$, we have

$$|G^k(0,0) - G^k(x,y)| > \varepsilon$$

Remark 1. Using similar procedures in a current work in progress, we are extending the constructions made on I^2 or $[0,1) \times [0,1)$ to I^n or $[0,1)^n$ for $n > 2$, using permutation maps introduced in [2].

References

1. Balibrea, F., Caballero, M. V.: Stability of orbits via Lyapunov exponents in autonomous and non-autonomous systems. International Journal of Bifurcation and chaos, 91, vol **23**, 1, 13501–13513 (2013)
2. Balibrea, F., Linero, A.: Periodic structure of σ-permutation maps on I^n. Aequ. Math. **61**, 1–15 (2001)
3. Demir, B., Koçak, S.: A note on positive Lyapunov exponent and sensitive dependence on initial conditions. Chaos, Soliton. Fract. **12**, 2119–2121 (2001)
4. Koçak, H., Palmer, K.: Lyapunov exponents and sensitive dependence. J.Dynam. Differ. Equat. **22**(3), 381–393 (2010)
5. Osedelets, I.: A multiplicative ergodic theorem. Lyapunov characteristic numbers for dynamical systems. Trans. Moscow Math. Soc. **19**, 197–221 (1968)

Chapter 3
Characteristics of the Explosive Percolation Transition

R.A. da Costa, S.N. Dorogovtsev, A.V. Goltsev, and J.F.F. Mendes

Abstract An unusual phase transition was recently discovered in a bunch of evolving network models in which each new connection between nodes is selected from several possibilities by an optimization algorithm. First studies (simulations) of these systems interpreted the emergence of a percolation cluster (giant connected component) as a discontinuous phase transition, and so the transition was said to be "explosive." We have shown, however, that this transition is actually continuous but with uniquely small critical exponent of the percolation cluster size. Here we propose an efficient method for finding the characteristics of this second-order transition for a set of explosive percolation models. For each of the models, with high precision, we obtain critical exponents and amplitudes, and the critical point.

3.1 Explosive Percolation Problem

Percolation problems are well known to show a continuous phase transition that in essence is the birth of a percolation cluster (a giant connected component, as it is called in graph theory) in a system [16]. This transition is observed for percolation on lattices and on various networks [6–9]. The simplest classical model for percolation (the classical random graph $G_{n,p}$ model) is formulated in the following way. Start from a large number N of isolated nodes (in fact, $N \to \infty$), and at each step

R.A. da Costa (✉) • J.F.F. Mendes
Department of Physics and I3N, University of Aveiro, Campus Universitário de Santiago, 3810-193 Aveiro, Portugal
e-mail: americo.costa@ua.pt; jfmendes@ua.pt

S.N. Dorogovtsev • A.V. Goltsev
Department of Physics and I3N, University of Aveiro, Campus Universitário de Santiago, 3810-193 Aveiro, Portugal

A.F. Ioffe Physico-Technical Institute, St. Petersburg, Russia
e-mail: sdorogov@ua.pt; goltsev@ua.pt

C. Grácio et al. (eds.), *Nonlinear Maps and their Applications*, Springer Proceedings in Mathematics & Statistics 57, DOI 10.1007/978-1-4614-9161-3_3,

choose uniformly at random a pair of nodes and connect them by a link. When the relative number of links $t = L/N$ in this graph exceeds the threshold $t_c = 1/2$, the graph has a giant cluster (in other words, a giant connected component or percolation cluster) containing a finite fraction $S = N_{GC}/N$ of all nodes. This process can be reversed, and so the percolation transition in this system is actually equilibrium. The continuity of the transition means that the order parameter S emerges continuously, without a jump at the critical point. In particular, for the classical random graph model described above, the relative size of the giant connected component S is proportional to the difference $(t - t_c)$ in a neighborhood of the percolation threshold, that is, the critical exponent β defined as $S \propto (t - t_c)^{\beta}$ is 1. In a neighborhood of the continuous phase transition, scaling behavior takes place. In particular, at the critical point, the cluster size distribution $n(s)$ decays asymptotically as a power law $n(s) \sim s^{-\tau}$, where the critical exponent τ is $5/2$ for the classical random graphs.

The continuity of percolation phase transitions was generally accepted until recently. This common understanding was shaken by the study [1], reporting a discontinuous percolation phase transition in models in which each new link is selected from several possibilities by a special optimization algorithm (e.g., of two attempted connections, the link merging two smallest clusters is established). The suggested discontinuity produced the impressive-sounding term "explosive percolation."

This observation was confirmed in a chain of subsequent works; see, e.g., [2–4, 13, 17]. One should note, however, that all these studies, including the first one, were based on simulations. In our work [5], we showed that the conclusions of our predecessors were incorrect, and the so-called explosive percolation transition is actually continuous. Importantly, we explained that the exponent β of this transition is surprisingly small, which makes the observation of the continuous transition in simulations of these systems virtually impossible. Afterward, the continuity of the explosive percolation transition was proved mathematically [14] and was observed in [12] for other models.

To describe quantitatively the explosive percolation transition, in our work [5] we reduced the problem to a specific aggregation process and solved numerically 10^6 evolution equations for the cluster size distribution ($s \leq 10^6$). The evolution equations for the explosive percolation problems resemble the Smoluchowski equation [15], which enables one to use standard numerical algorithms [11]. In the present work, we demonstrate how to find characteristics of the explosive percolation transition with even higher precision by implementing a simpler method, without solving so large an array of evolution equations.

3.2 The Model and Evolution Equations

Here we use the version of the explosive percolation process proposed in [5]. This representative and elegant model naturally generalizes the classical random graph model of ordinary percolation (see Sect. 3.1). Starting from a given set of clusters,

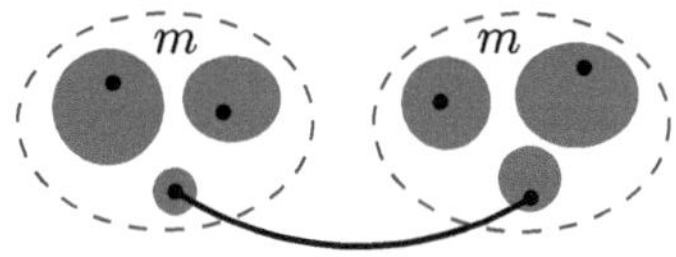

Fig. 3.1 Schematic representation of the explosive percolation model rules. At each step, two samples of m nodes are selected uniformly at random. In each of the samples, the node belonging to the smallest (of m) clusters is chosen, and a new link connecting these two nodes is added to the system. In other words, two sets of clusters are chosen with probability proportional to their sizes, and the two smallest clusters, taken from each of the sets, merge

e.g., from a large number N of isolated nodes, at each step, we do the following; see Fig. 3.1. We select uniformly at random m nodes and choose the one that is inside the smallest of the clusters to which these nodes belong. Then we again select m nodes and choose the one belonging to the smallest cluster, and finally, we add an edge connecting the two nodes selected in this way. If, in particular, $m = 1$, we recover the classical random graph model. Here we consider the cases $m = 2$, 3, and 4.

This model can be treated as an aggregation process, in which clusters selected by our rules merge progressively. A complete description of this process is provided by the evolving distribution $P(s,t) = sn(s)/\langle s \rangle$, which is the probability that a uniformly randomly chosen node belongs to a cluster of size s at time t. Here $n(s)$ is the size distribution of clusters and $\langle s \rangle$ is the average size of clusters (including the percolation cluster). Another important characteristic is the probability $Q(s,t)$ that the node chosen by the rules formulated above from an m-node sample belongs to a cluster of size s. In other words, this is the size distribution of merging clusters. Using basic formulas of probability theory [10], one can express the distribution $Q(s,t)$ in terms of $P(s,t)$, namely

$$Q(s,t) \equiv \left(\sum_{u=s}^{\infty} P(u,t) \right)^m - \left(\sum_{u=s+1}^{\infty} P(u,t) \right)^m . \tag{3.1}$$

The evolution equations for the distributions $P(s,t)$ and $Q(s,t)$ describing this aggregation process in the infinite system ($N \to \infty$) look as follows:

$$\frac{\partial P(s,t)}{\partial t} = s \sum_{u+v=s} Q(u,t)Q(v,t) - 2sQ(s,t) . \tag{3.2}$$

In particular, at $m = 1$, $Q(s,t)$ in Eq. (3.2) should be substituted for the distribution $P(s,t)$, and we arrive at the well-known master equations for standard percolation, which can be easily solved by implementing a standard generating function technique. This cannot be done in general, for arbitrary m. We should analyze the difficult Eq. (3.2), taking into account the relation (3.1) and a given initial distribution $P(s,0)$. The solution of this equation has the power-law asymptotics

$$P(s,t_c) \cong f(0)s^{1-\tau} \tag{3.3}$$

at the critical point. Here $f(0)$ is a critical amplitude, which coincides with the value of the scaling function $f(x)$ at $x = 0$. Near t_c we have $P(s,t) = s^{1-\tau} f\left(s|t-t_c|^{1/\sigma}\right)$, where τ and σ are critical exponents. Note that while the values of the critical exponents are independent of the initial cluster size distributions (initial distributions should decay sufficiently rapidly to produce a nonzero critical point $t_c > 0$), the form of the scaling function $f(x)$ (and so the critical amplitude $f(0)$) depends on the initial distribution $P(s,0)$. Furthermore, the scaling function above the critical point differs from that below t_c. One can show that the critical exponents of this problem can be expressed in terms of one of them, and so we have to find only one critical exponent. In our work [5], we have derived these relations in the particular case $m = 2$. Similarly, for arbitrary m, we find the following relation between the critical exponents τ and β:

$$\beta = \frac{\tau - 2}{1 - (2m-1)(\tau - 2)} \,. \tag{3.4}$$

Thus our main aim is to find one of the critical exponents, say τ, and the critical amplitude $f(0)$ for $m = 2$, 3, and 4. For the sake of brevity, here we assume that at the initial moment all the nodes are disconnected, that is, $P(s,0) = \delta_{s,1}$, where $\delta_{s,1}$ is the Kronecker symbol.

Conveniently, Eq. (3.2) is a chain of coupled equations, in which the first equation gives $P(1,t)$; substituting this result into the second equation and solving it gives $P(2,t)$, and so on. One can solve numerically first s_0 equations and find $P(s \leq s_0, t)$ at any t with any desired precision. In our work [5], we fulfilled this program, directly solved the evolution equations, and in particular, found the distribution $P(s \leq 10^6, t_c)$ at the critical point in the case of $m = 2$. Let us use a more efficient technique to obtain the characteristics of the explosive percolation transition.

3.3 The Method

Here we briefly describe the idea of our method. A more detailed explanation will be presented elsewhere. At the critical point, the distribution $P(s,t_c)$ has the power-law asymptotics (3.3). We can solve numerically a finite number of equations (3.2), say s_{m}. This provides us with $P(s,t)$ for $s < s_{\mathrm{m}}$ and any time $t \leq 1$.

If we knew the value of the critical point t_c, we could readily obtain the exponent τ and the critical amplitude $f(0)$ by sewing together the distribution $P(s \leq s_{\mathrm{m}}, t_c)$ obtained by numerical solution of the evolution equations and the power-law asymptotics $P(s \geq s_{\mathrm{m}}, t_c) = f(0)s^{1-\tau}$. For this, we should use two conditions: (1) $P(s_{\mathrm{m}}, t_c) = f(0)s^{1-\tau}$ and (2) normalization, namely $\sum_{s=1}^{\infty} P(s,t_c) = 1$. We would get a precise result in the limit $s_{\mathrm{m}} \to \infty$.

Although we do not know the value of t_c in advance, we can formally perform this procedure at t deviating from the critical point. The results obtained at two values of t, smaller and greater than t_c, are shown in Fig. 3.2 in the case $m = 2$. Here, the solid curve shows the set of points $\tau(s_{\mathrm{m}})$ at some $t < t_c$ (normal phase),

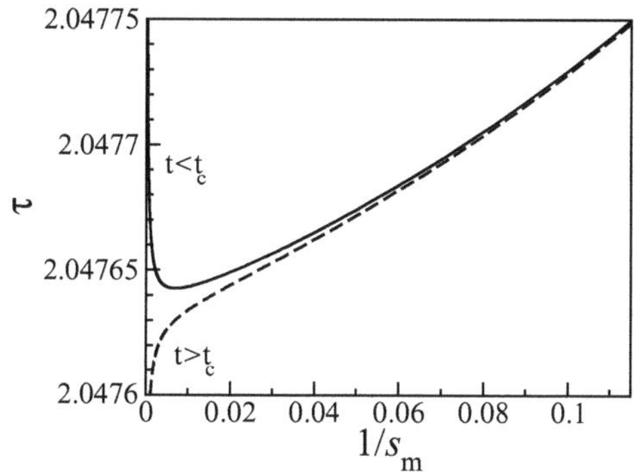

Fig. 3.2 Variation of the calculated τ with $1/s_{\mathrm{m}}$, where s_{m} denotes the maximum cluster size taken into account in the evolution equations, $m = 2$. In our calculations, s_{m} takes a large set of values from 2 to 10^5. The solid and dashed curves correspond to two values of t, namely $t < t_c$ and $t > t_c$, respectively

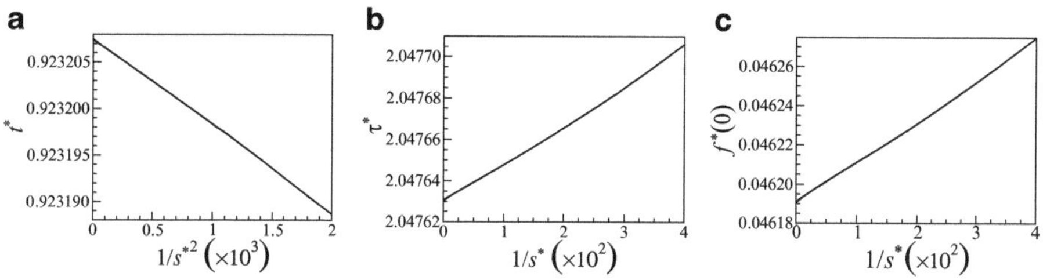

Fig. 3.3 The extrapolation of (**a**) t^*, (**b**) τ^*, and (**c**) $f^*(0)$ corresponding to the minima $1/s^*$ of the $\tau(1/s_{\mathrm{m}}, t < t_c)$ curves in Fig. 3.2 to the precise values t_c, τ, and $f(0)$, respectively, at $s^* \to \infty$ in the case $m = 2$

and the dashed curve shows the set of points $\tau(s_{\mathrm{m}})$ at some $t > t_c$, which corresponds to a phase with a giant component. The plot shows how the value of the exponent τ, found using the sewing procedure, varies with the inverse maximum cluster size $1/s_{\mathrm{m}}$ taken into account in the equations. With diminishing $1/s_{\mathrm{m}}$, these curves approach infinity if t is below t_c, and 2 if it is above. Here the infinite exponent τ corresponds to a rapidly decreasing cluster size distribution in the normal phase. On the other hand, the value 2 in this indicates the failure of the power-law form, which we assumed, to describe correctly the distribution of clusters in the phase with a giant component. If we guessed t_c properly, the curve would lead to the precise τ as $1/s_{\mathrm{m}} \to 0$. Otherwise, the curves run away from that point, which is precisely the behavior demonstrated by the solid and dashed curves in Fig. 3.2. To find the precise values of t_c, τ, and $f(0)$, we analyzed how the point of the minimum $1/s^*$ of the solid curve $\tau(1/s_{\mathrm{m}})$ in Fig. 3.2 relates to the values t^*, τ^*, and $f^*(0)$ corresponding to this minimum. Figure 3.3 demonstrates that t^* approaches t_c linearly with $1/s^{*2}$, and τ^* and $f(0)^*$ approach respectively τ and $f(0)$ linearly with $1/s^*$. This enables

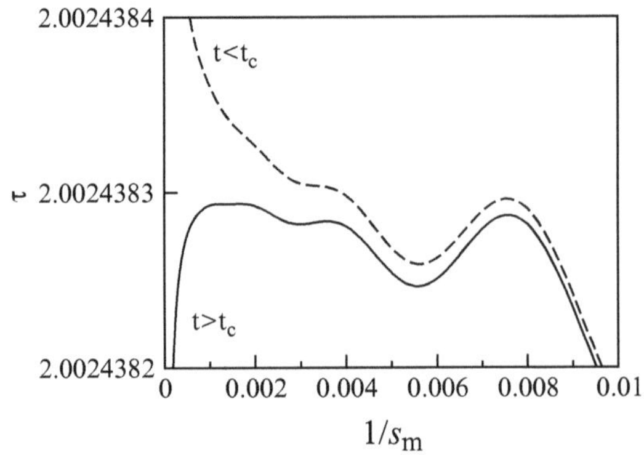

Fig. 3.4 Variation of the calculated τ with $1/s_{\mathrm{m}}$ in the case of $m = 4$. The *solid* and *dashed curves* correspond to two values of t, namely $t > t_c$ and $t < t_c$, respectively

us to extrapolate to $s^* \to \infty$ (the maximum s_{m}, that is, the maximum number of equations that we solved numerically was 10^5) and find t_c, τ, and $f(0)$ with a very high precision.

In principle, one can even avoid extrapolation procedures, which may turn out to be difficult at $m = 4$ and higher. The difficulty is that the curves $\tau(1/s_{\mathrm{m}})$ oscillate (see Fig. 3.4), because the distribution $P(s,t)$ oscillates in the range of low s. The reason for these oscillations is the following. If m tends to infinity, then according to our rules, two smallest clusters in the system should merge at each step. This is why first single nodes merge into the clusters of size 2, then these clusters merge into the clusters of four nodes, and so on. This results in the peaks of the distribution at $s = 2, 4, 8, \ldots$, which are seen already at $m = 4$. In turn, these peaks lead to the oscillations in Fig. 3.4. Fortunately, the amplitude of these oscillations decreases with decreasing $1/s_{\mathrm{m}}$. So one can easily study how the curves run away from the precise value of τ at small $1/s_{\mathrm{m}}$ and find t_c by adjusting t in such a way that the runaway occurs at the smallest value of $1/s_{\mathrm{m}}$.

3.4 Characteristics of the Transition

The results of the application of this method to the models with $m = 2$, 3, and 4 are shown in Table 3.1. For comparison, in the first column of the table, we show the exact values for the normal percolation problem ($m = 1$). The values of the exponent β are found from τ using the relation (3.4). In the case of $m = 2$, the values presented in the table agree with our results [5], although the precision of the numbers obtained in the present work is much higher. Furthermore, the results in the table for the models with $2 \leq m \leq 4$ agree with those obtained from equations for scaling functions (we will consider this alternative method elsewhere). The table shows that as is natural, as m increases, the difference $1 - t_c$ diminishes, and the exponent β of

Table 3.1 Characteristics of the standard percolation ($m = 1$) and explosive percolation ($m = 2,3,4$) transitions

m	1	2	3	4
t_c	$1/2$	0.923207509297(2)	0.9817953173509(2)	0.99497356260563(2)
β	1	0.05557108(1)	0.010428725(1)	0.0024806708
τ	$5/2$	2.04763045(1)	2.009911883(1)	2.0024383299(1)
$f(0)$	$1/\sqrt{2\pi}\approx 0.3989$	0.04619071(1)	0.009831398(1)	0.0024320386(1)
$P(1,t_c)$	$1/e\approx 0.3678$	0.0485928295546(4)	0.01172146480245(2)	0.003343067143133(1)

the giant component size also decreases. One should note that the critical amplitude $f(0)$ is close to $P(1,t_c)$, especially when $m \leq 3$. This closeness indicates that the deviations from power-law asymptotics in these problems are small even at low s. Interestingly, $f(0) > P(1,t_c)$ for classical percolation, while the opposite is true for the explosive percolation transition. The values of β are remarkably small. In particular, in the case of $m = 4$, β is close to $1/400$. In simulating these systems, the smallest step in t corresponds to addition of a single link and equals $1/N$, that is, the inverse size of the system. Let m be 4. Then the minimum jump at t_c, which one can observe in simulating even an unrealistically large system of 10^{400} nodes, is of order $(10^{-400})^{1/400} = 10^{-1}$, which makes observation of a continuous transition in simulations impossible. We suggest that finite-size effects in this situation can hardly be investigated. Even in the case of $m = 2$, in which β is close to $1/18$, in simulating a system of 10^{18}, one cannot observe a jump smaller than of order 10^{-1} at the critical point.

3.5 Conclusions

We have proposed an effective numerical method enabling us to find basic characteristics of explosive percolation transitions with high precision. We obtained the critical points, and the critical exponents and critical amplitudes of scaling functions of these transitions in a set of representative models of explosive percolation. Our results show that in all these models, the explosive percolation transitions are continuous. Moreover, we conclude that in a wide range of models with rules similar to those described in Sects. 3.1 and 3.2, the transition is continuous. In particular, these systems include the models considered in [1, 12]. Our approach provides a useful tool for a quantitative description of a new class of critical phenomena in nonequilibrium systems and irreversible processes.

Acknowledgements This work was partially supported by the FCT projects PTDC: FIS/71551/2006, FIS/108476/2008, SAU-NEU/103904/2008, MAT/114515/2009, and PEst-C/CTM/LA0025/2011.

References

1. Achlioptas, D., D'Souza, R.M., Spencer, J.: Explosive percolation in random networks. Science **323**, 1453–1455 (2009)
2. Araujo, N.A.M., Andrade Jr., J.S., Ziff, R.M., Herrmann, H.J.: Tricritical point in explosive percolation. Phys. Rev. Lett. **106**, 095703 (2011)
3. Cho, Y.S., Kim, J.S., Park, J., Kahng, B., Kim, D.: Percolation transitions in scale-free networks under Achlioptas process. Phys. Rev. Lett. **103**, 135702 (2009)
4. Cho, Y.S., Kahng, B., Kim, D.: Cluster aggregation model for discontinuous percolation transition. Phys. Rev. E **81**, 030103(R) (2010)
5. da Costa, R.A., Dorogovtsev, S.N., Goltsev, A.V., Mendes, J.F.F.: Explosive percolation transition is actually continuous. Phys. Rev. Lett. **105**, 255701. http://arxiv.org/abs/1009.2534 (2010)
6. Dorogovtsev, S.N.: Lectures on Complex Networks. Oxford University Press, Oxford (2010)
7. Dorogovtsev, S.N., Mendes, J.F.F.: Evolution of networks. Adv. Phys. **51**, 1079 (2002)
8. Dorogovtsev, S.N., Mendes, J.F.F.: Evolution of Networks: From Biological Nets to the Internet and WWW. Oxford University Press, Oxford (2003)
9. Dorogovtsev, S.N., Goltsev, A.V., Mendes, J.F.F.: Critical phenomena in complex networks. Rev. Mod. Phys. **80**, 1275 (2008)
10. Feller, W.: An Introduction to Probability Theory and Its Applications, vol. 1 and 2. Wiley, New York (1968)
11. Leyvraz, F., Tschudi, H.R.: Singularities in the kinetics of coagulation processes. J. Phys. A: Math. Gen. **14**, 3389 (1981)
12. Nagler, J., Levina, A., Timme, M.: Impact of single links in competitive percolation. Nature Phys. **7**, 265 (2011)
13. Radicchi, F., Fortunato, S.: Explosive percolation in scale-free networks. Phys. Rev. Lett. **103**, 168701 (2009)
14. Riordan, O., Warnke, L.: Achlioptas process phase transitions are continuous. Science **333**, 322 (2011)
15. Smoluchowski, M.V.: Über Brownsche Molekularbewegung unter Einwirkung äußerer Kräfte und den Zusammenhang mit der verallgemeinerten Diffusionsgleichung. Ann. Phys. **353**, 1103 (1915)
16. Stauffer, D., Aharony, A.: Introduction to Percolation Theory. Taylor & Francis, London (1994)
17. Ziff, R.M.: Explosive growth in biased dynamic percolation on two-dimensional regular lattice networks. Phys. Rev. Lett. **103**, 045701 (2009)

Chapter 4
Controlled Markovian Dynamics of Graphs: Unbiased Generation of Random Graphs with Prescribed Topological Properties

E.S. Roberts, A. Annibale, and Anthony C.C. Coolen

Abstract We analyze the properties of degree-preserving Markov chains based on elementary edge switchings in undirected and directed graphs. We give exact yet simple formulas for the mobility of a graph (the number of possible moves) in terms of its adjacency matrix. This formula allows us to define acceptance probabilities for edge switchings such that the Markov chains become controlled Glauber-type detailed balance processes, designed to evolve to any required invariant measure (representing the asymptotic frequencies with which the allowed graphs are visited during the process).

4.1 Introduction

Sampling uniformly the space of graphs with prescribed macroscopic properties has become a prominent problem in many application areas where tailored graph ensembles are used as proxies or null models for real complex networks. Quantities measured in real networks are often compared with the values these quantities take in their randomized counterparts. Ensembles of randomized networks allow one to put error bars on these values, and therefore to identify which topological features

E.S. Roberts (✉)
Randall Division of Cell and Molecular Biophysics, King's College, London, UK
e-mail: ekaterina.roberts@kcl.ac.uk

A. Annibale
Department of Mathematics, King's College, London, UK
e-mail: alessia.annibale@kcl.ac.uk

A.C.C. Coolen
Department of Mathematics and Randall Division of Cell and Molecular Biophysics, King's College, London, UK
e-mail: ton.coolen@kcl.ac.uk

C. Grácio et al. (eds.), *Nonlinear Maps and their Applications*, Springer Proceedings in Mathematics & Statistics 57, DOI 10.1007/978-1-4614-9161-3_4,

of real networks deviate significantly from the null model. Such features are likely to reflect such aspects of the network as design principles and evolutionary history.

Ensembles of randomized graphs are to be generated numerically, and each graph realization should be produced with a probability proportional to a prescribed statistical weight (often taken to be uniform) of the graph for the analysis to be unbiased. At the level where the constraints of the randomized graphs involve only the simplest quantities, such as the degree sequence, even uniform generation of such graphs is known to be a nontrivial problem [1, 2, 6, 11].

One classical algorithm for generating random networks with prescribed degrees [7] assigns to each node a number of "edge stubs" equal to its desired degree and joins iteratively pairs of randomly chosen stubs to form a link. A drawback of this algorithm is the need for rejection of forbidden graphs (those with multiple edges or self-loops), which can lead to biased sampling [5]. A second popular method for generating random graphs with a given degree sequence is "edge swapping," which involves successions of ergodic graph randomizing moves that leave all degrees invariant [4, 12].

However, naive accept-all edge swapping will again cause sampling biases. The reason is that the *number* of edge swaps that can be executed is not a constant; it depends on the graph $\mathbf{c}$ at hand; graphs that allow for many moves will be generated more often. Any bias in the sampling of graphs invalidates their use as null models, so one is forced to mistrust all papers in which observations in real graphs have been tested against null models generated either via the "stubs" method or via randomization by "accept-all edge swapping."

This situation can be remedied via a systematic study of stochastic Markovian graph dynamics, which is the topic of this paper. We determine the appropriate adjustment to the probability of accepting a randomly chosen proposed edge swap for the process to visit each graph configuration with the same probability. This can be done for undirected and directed graphs. We will also show that our method can be used to generate graphs with prescribed degree correlations.

4.2 Generating Random Graphs via Markov Chains

A general and exact method for generating graphs from the set $G[\mathbf{k}] = \{\mathbf{c} \in G|\ \mathbf{k}(\mathbf{c}) = \mathbf{k}\}$ randomly (where $\mathbf{k}$ denotes the degree sequence, or the joint in- and out-degree sequence of the graph) with specified probabilities $p(\mathbf{c}) = Z^{-1} \exp[-H(\mathbf{c})]$ was developed in [3]. It has the form of a Markov chain:

$$\forall \mathbf{c} \in G[\mathbf{k}]: \quad p_{t+1}(\mathbf{c}) = \sum_{\mathbf{c}' \in G[\mathbf{k}]} W(\mathbf{c}|\mathbf{c}') p_t(\mathbf{c}'). \tag{4.1}$$

Here $p_t(\mathbf{c})$ is the probability of observing graph $\mathbf{c}$ at time t in the process, and $W(\mathbf{c}|\mathbf{c}')$ is the one-step transition probability from graph $\mathbf{c}'$ to $\mathbf{c}$. For any set Φ of *ergodic* reversible elementary moves $F: G[\mathbf{k}] \to G[\mathbf{k}]$, we can choose transition probabilities of the form

$$W(\mathbf{c}|\mathbf{c}') = \sum_{F\in\Phi} q(F|\mathbf{c}')\Big[\delta_{\mathbf{c},F\mathbf{c}'}A(F\mathbf{c}'|\mathbf{c}') + \delta_{\mathbf{c},\mathbf{c}'}[1-A(F\mathbf{c}'|\mathbf{c}')]\Big]. \tag{4.2}$$

The interpretation is as follows. At each step, a candidate move $F \in \Phi$ is drawn with probability $q(F|\mathbf{c}')$, where $\mathbf{c}'$ denotes the current graph. This move is accepted (and the move $\mathbf{c}' \to \mathbf{c} = F\mathbf{c}'$ executed) with probability $A(F\mathbf{c}'|\mathbf{c}') \in [0,1]$, which depends on the current graph $\mathbf{c}'$ and the proposed new graph $F\mathbf{c}'$. If the move is rejected, which happens with probability $1-A(F\mathbf{c}'|\mathbf{c}')$, the system stays in $\mathbf{c}'$. We may always exclude from Φ the identity operation.

One can prove that the process (4.1) will converge to the equilibrium measure $p_\infty(\mathbf{c}) = Z^{-1}\exp[-H(\mathbf{c})]$ on making in (4.2) the choices

$$q(F|\mathbf{c}) = I_F(\mathbf{c})/n(\mathbf{c}), \tag{4.3}$$

$$A(\mathbf{c}|\mathbf{c}') = \frac{n(\mathbf{c}')e^{-\frac{1}{2}[H(\mathbf{c})-H(\mathbf{c}')]}}{n(\mathbf{c}')e^{-\frac{1}{2}[H(\mathbf{c})-H(\mathbf{c}')]} + n(\mathbf{c})e^{\frac{1}{2}[H(\mathbf{c})-H(\mathbf{c}')]}}. \tag{4.4}$$

Here $I_F(\mathbf{c}) = 1$ if the move F can act on graph $\mathbf{c}$, $I_F(\mathbf{c}) = 0$ otherwise, and $n(\mathbf{c})$ denotes the total number of moves that can act on a graph $\mathbf{c}$ (the "mobility" of state $\mathbf{c}$):

$$n(\mathbf{c}) = \sum_{F\in\Phi} I_F(\mathbf{c}). \tag{4.5}$$

4.3 Degree-Constrained Dynamics of Undirected Graphs

We first apply our results to algorithms that randomize undirected graphs while conserving all degrees by application of edge swaps that act on quadruplets of nodes and their mutual links. Such moves were shown to be ergodic, i.e., every pair of graphs with the same degree sequence can be connected by a finite number of successive edge swaps [4, 12].

Let us define the set $Q = \{(i,j,k,\ell) \in \{1,\ldots,N\}^4 |\, i<j<k<\ell\}$ of all ordered node quadruplets. The possible edge swaps to act on (i,j,k,ℓ) are the following, with thick lines indicating existing links and thin lines indicating absent links that will be swapped with the existing ones, and where (IV, V, VI) are the inverses of (I, II, III):

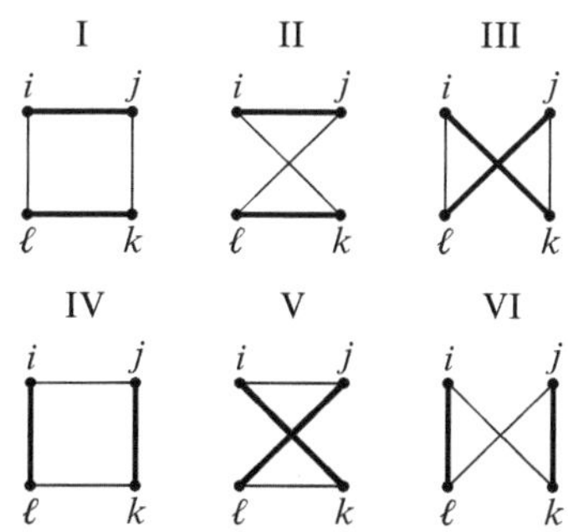

We group the edge swaps into three pairs, (I, IV), (II, V), and (III, VI), and label all three resulting autoinvertible operations for each ordered quadruple (i,j,k,ℓ) with a subscript α. Our autoinvertible edge swaps are now written as $F_{ijk\ell;\alpha}$, with $i<j<k<\ell$ and $\alpha \in \{1,2,3\}$. We define associated indicator functions $I_{ijk\ell;\alpha}(\mathbf{c}) \in \{0,1\}$ that detect whether (1) or not (0) the edge swap $F_{ijk\ell;\alpha}$ can act on state $\mathbf{c}$, so

$$I_{ijk\ell;1}(\mathbf{c}) = c_{ij}c_{k\ell}(1-c_{i\ell})(1-c_{kj}) + (1-c_{ij})(1-c_{k\ell})c_{i\ell}c_{kj}, \tag{4.6}$$
$$I_{ijk\ell;2}(\mathbf{c}) = c_{ij}c_{\ell k}(1-c_{ik})(1-c_{\ell j}) + (1-c_{ij})(1-c_{\ell k})c_{ik}c_{\ell j}, \tag{4.7}$$
$$I_{ijk\ell;3}(\mathbf{c}) = c_{ik}c_{j\ell}(1-c_{i\ell})(1-c_{jk}) + (1-c_{ik})(1-c_{j\ell})c_{i\ell}c_{jk}. \tag{4.8}$$

If $I_{ijk\ell;\alpha}(\mathbf{c}) = 1$, this edge swap will operate as follows:

$$F_{ijk\ell;\alpha}(\mathbf{c})_{qr} = 1 - c_{qr} \quad \text{for } (q,r) \in \mathcal{S}_{ijk\ell;\alpha}, \tag{4.9}$$
$$F_{ijk\ell;\alpha}(\mathbf{c})_{qr} = c_{qr} \quad \text{for } (q,r) \notin \mathcal{S}_{ijk\ell;\alpha}, \tag{4.10}$$

where

$$\mathcal{S}_{ijk\ell;1} = \{(i,j),(k,\ell),(i,\ell),(k,j)\}, \tag{4.11}$$
$$\mathcal{S}_{ijk\ell;2} = \{(i,j),(\ell,k),(i,k),(\ell,j)\}, \tag{4.12}$$
$$\mathcal{S}_{ijk\ell;3} = \{(i,k),(j,\ell),(i,\ell),(j,k)\}. \tag{4.13}$$

Insertion of these definitions into the recipe (4.2)–(4.4) gives

$$W(\mathbf{c}|\mathbf{c}') = \sum_{i<j<k<\ell} \sum_{\alpha \le 3} \frac{I_{ijk\ell;\alpha}(\mathbf{c}')}{n(\mathbf{c}')} \times \left[\frac{\delta_{\mathbf{c},F_{ijk\ell;\alpha}\mathbf{c}'} \mathrm{e}^{-\frac{1}{2}[E(F_{ijk\ell;\alpha}\mathbf{c}')-E(\mathbf{c}')]} + \delta_{\mathbf{c},\mathbf{c}'} \mathrm{e}^{\frac{1}{2}[E(F_{ijk\ell;\alpha}\mathbf{c}')-E(\mathbf{c}')]}}{\mathrm{e}^{-\frac{1}{2}[E(F_{ijk\ell;\alpha}\mathbf{c}')-E(\mathbf{c}')]} + \mathrm{e}^{\frac{1}{2}[E(F_{ijk\ell;\alpha}\mathbf{c}')-E(\mathbf{c}')]}} \right], \tag{4.14}$$

with $E(\mathbf{c}) = H(\mathbf{c}) + \log n(\mathbf{c})$. The process (4.14) can be described as the following algorithm. Given an instantaneous graph $\mathbf{c}'$, (i) choose uniformly at random a quadruplet (i,j,k,ℓ) of sites; (ii) if at least one of the three edge swaps $\mathbf{c}' \to \mathbf{c} = F_{ijk\ell;\alpha}(\mathbf{c}')$ is possible, select one of these uniformly at random and execute it with an acceptance probability

$$A(\mathbf{c}|\mathbf{c}') = \left[1 + \mathrm{e}^{E(F_{ijk\ell;\alpha}\mathbf{c}')-E(\mathbf{c}')}\right]^{-1}; \tag{4.15}$$

then return to (i). For this Markov chain recipe to be practical, we finally need a formula for the mobility $n(\mathbf{c})$ of a graph. This could be calculated [3], giving[1] (with $\mathrm{Tr}\mathbf{A} = \sum_i A_{ii}$)

$$n(\mathbf{c}) = \frac{1}{4}\Big(\sum_i k_i\Big)^2 + \frac{1}{4}\sum_i k_i - \frac{1}{2}\sum_i k_i^2 - \frac{1}{2}\sum_{ij} k_i c_{ij} k_j + \frac{1}{4}\mathrm{Tr}(\mathbf{c}^4) + \frac{1}{2}\mathrm{Tr}(\mathbf{c}^3). \tag{4.16}$$

[1] Efficient expressions for the change in mobility following one move have also been derived [3] to avoid unnecessary matrix multiplication in the computational implementation.

Naive "accept-all" edge swapping, where $A(\mathbf{c}|\mathbf{c}') = 1$, corresponds to choosing $E(\mathbf{c}) = 0$ in (4.14), and it would give the *biased* graph-sampling probabilities $p_\infty(\mathbf{c}) = n(\mathbf{c})/\sum_{\mathbf{c}'} n(\mathbf{c}')$ on equilibration.

The graph mobility acts as an entropic force, which can be neglected only if (4.16) is dominated by its first three terms; it has been shown that a sufficient condition for this to be true is $\langle k^2\rangle k_{\max}/\langle k\rangle^2 \ll N$. This condition holds in networks with narrow degree sequences, and naive edge swapping is roughly acceptable. However, one has to be careful with scale-free graphs, where $\langle k^2\rangle$ and $k_{\max}$ diverge as $N \to \infty$.

4.4 Degree-Constrained Dynamics of Directed Graphs

For directed networks, the generalization of the canonical edge swap (up to relabeling of nodes) is shown below:

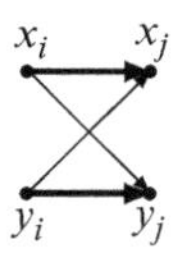

This move samples the space of all directed graphs with prescribed in- and out-degrees ergodically only if self-interactions are permitted [9]. We could now proceed with our formalism as before, but it is more efficient in the case of directed graphs to define the swaps and the indicator function in terms of pairs of links rather than quadruples of nodes.

Let $\mathbf{c}$ now be an asymmetric connectivity matrix such that $c_{ab} = 1$ if $a \to b$, otherwise $c_{ab} = 0$. Consider Λ to be the set of links within the network defined by $\mathbf{c}$. We write $\mathbf{x} = (x_i, x_j) \in \Lambda$ if and only if $c_{x_i x_j} = 1$. The indicator function can be written as

$$I_{\mathbf{x},\mathbf{y};\Box} = \begin{cases} 1 & \text{if } \mathbf{x},\mathbf{y} \in \Lambda \text{ and } (x_i,y_j),(y_i,x_j) \notin \Lambda \\ 0 & \text{otherwise} \end{cases}$$

If $I_{\mathbf{x},\mathbf{y};\Box} = 1$, then the corresponding autoinvertible operation $F_{\mathbf{x},\mathbf{y};\Box}$ acts on state $\mathbf{c}$ as follows:

$$F_{\mathbf{x},\mathbf{y};\Box}(\mathbf{c})_{qr} = 1 - c_{qr} \quad \text{for } q \in \{x_i, y_i\} \text{ and } r \in \{x_j, y_j\}, \tag{4.17}$$
$$F_{\mathbf{x},\mathbf{y};\Box}(\mathbf{c})_{qr} = c_{qr} \quad \text{otherwise.} \tag{4.18}$$

When self-interactions are forbidden, a further type of elementary move is required to ensure ergodicity [9]. It can be visualized as reversing a triangle cycle:

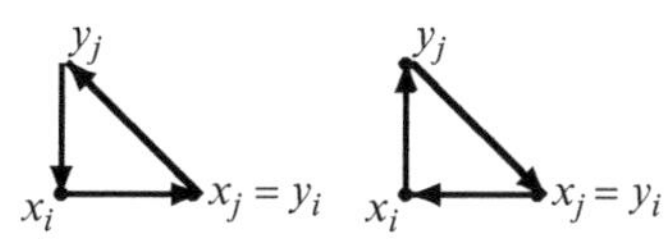

The indicator function for this new move is

$$I_{\mathbf{x},\mathbf{y};\triangle} = \begin{cases} 1 & \text{if } \mathbf{x},\mathbf{y},(y_j,x_i) \in \Lambda \text{ and } x_j = y_i \text{ and } \mathbf{x}^{-1},\mathbf{y}^{-1},(x_i,y_j) \notin \Lambda, \\ 0 & \text{otherwise.} \end{cases}$$

where $\mathbf{x}^{-1} = (x_j, x_i)$ represents a link that is to be reversed. The corresponding autoinvertible operation represented by $F_{\mathbf{x},\mathbf{y};\triangle}$ acts on state $\mathbf{c}$ as follows:

$$F_{\mathbf{x},\mathbf{y};\triangle}(\mathbf{c})_{qr} = 1 - c_{qr} \quad \text{for } (q,r) \in \mathcal{S}_{x_i,x_j,y_j}, \tag{4.19}$$

$$F_{\mathbf{x},\mathbf{y};\triangle}(\mathbf{c})_{qr} = c_{qr} \quad \text{for } (q,r) \notin \mathcal{S}_{x_i,x_j,y_j}, \tag{4.20}$$

in which $\mathcal{S}_{abc} = \{(a,b),(b,c),(c,a),(b,a),(c,b),(a,c)\}$ is the set of pairs of vertices of the triangle.

We now combine edge swaps and triangular reversions into (4.2). This requires us to express $\sum_{F\in\Phi} q(F|\mathbf{c}')$ from the point of view of sampling links $\mathbf{x}$ and $\mathbf{y}$. This is, in fact, straightforward, since each pair of links can trigger the square indicator function (if they define an edge swap), a triangle indicator function (if they define a triangle reversion), or neither (when the indicator function returns zero). It remains only to observe the symmetries of the problem: cycling through each pair of bonds, we will come across each unique edge swap twice, and across each unique triangle swap three times [10]. Hence

$$\sum_{F\in\Phi} q(F|\mathbf{c}') = \sum_{\mathbf{x},\mathbf{y}\in\Lambda} \frac{\frac{1}{2}I_{\mathbf{x},\mathbf{y};\square} + \frac{1}{3}I_{\mathbf{x},\mathbf{y};\triangle}}{n(\mathbf{c}')},$$

which expresses sampling over all possible moves in terms of (the more tangible) sampling over links. Insertion of the above definitions in (4.2)–(4.4) gives

$$W(\mathbf{c}|\mathbf{c}') = \sum_{\mathbf{x},\mathbf{y}\in\Lambda} \frac{\frac{1}{2}I_{\mathbf{x},\mathbf{y};\square} + \frac{1}{3}I_{\mathbf{x},\mathbf{y};\triangle}}{n(\mathbf{c}')} \times \left[\frac{\delta_{\mathbf{c},F_{\mathbf{x},\mathbf{y}}\mathbf{c}'} \mathrm{e}^{-\frac{1}{2}[E(F_{\mathbf{x},\mathbf{y}}\mathbf{c}')-E(\mathbf{c}')]} + \delta_{\mathbf{c},\mathbf{c}'} \mathrm{e}^{\frac{1}{2}[E(F_{\mathbf{x},\mathbf{y}}\mathbf{c}')-E(\mathbf{c}')]}}{\mathrm{e}^{-\frac{1}{2}[E(F_{\mathbf{x},\mathbf{y}}\mathbf{c}')-E(\mathbf{c}')]} + \mathrm{e}^{\frac{1}{2}[E(F_{\mathbf{x},\mathbf{y}}\mathbf{c}')-E(\mathbf{c}')]}}\right], \tag{4.21}$$

where $n(\mathbf{c}) = n_\square(\mathbf{c}) + n_\triangle(\mathbf{c})$ is the total number of valid moves that can act on the directed graph $\mathbf{c}$, written as the sum of the number of edge swaps and the number of triangle reversions. The mobility terms can again be calculated [10], giving

$$n_\square(\mathbf{c}) = \frac{1}{2}Tr(\mathbf{c}\mathbf{c}^T\mathbf{c}\mathbf{c}^T) - \sum_{i,j} \mathbf{k}_i^{out} c_{ij} \mathbf{k}_j^{in} + Tr(\mathbf{c}\mathbf{c}^T\mathbf{c}) + \frac{1}{2}(\sum_i \mathbf{k}_i)^2 + \frac{1}{2}Tr(\mathbf{c}^2) - \sum_i \mathbf{k}_j^{out}\mathbf{k}_j^{in},$$

$$n_\triangle(\mathbf{c}) = \frac{1}{3}Tr(\mathbf{c}^3) - Tr(\mathbf{c}^{\updownarrow}\mathbf{c}^2) + Tr(\mathbf{c}^{\updownarrow^2}\mathbf{c}) - \frac{1}{3}Tr(\mathbf{c}^{\updownarrow^3}),$$

with $(\mathbf{c}^T)_{ij} = c_{ji}$ and $(\mathbf{c}^{\updownarrow})_{ij} = c_{ij}c_{ji}$. Also for directed graphs it is possible to calculate efficient expressions for how the mobilities change following a single move.

For the simplest case, in which $H(\mathbf{c})$ is constant, i.e., the Markov chain is to evolve toward a flat measure, we immediately observe that the required move acceptance probabilities are

$$A(\mathbf{c}|\mathbf{c}') = \Big[1 + \frac{n_\square(\mathbf{c}) + n_\triangle(\mathbf{c})}{n_\square(\mathbf{c}') + n_\triangle(\mathbf{c}')}\Big]^{-1}. \tag{4.22}$$

4.5 Numerical Examples for Directed Graphs

Examples illustrating the canonical Markov chains for undirected graphs are given in [3]. Here we compare two variants of the *directed* edge-rewiring algorithm: the naive "accept all moves" version and the version with the canonical mobility corrections (4.22). Consider a network with $K+2$ nodes, one of which has degrees $(k^{\text{in}}, k^{\text{out}}) = (0, K)$, one has $(k^{\text{in}}, k^{\text{out}}) = (K, 0)$, and the remaining K have degrees $(k^{\text{in}}, k^{\text{out}}) = (1, 1)$. Self-interactions are forbidden. Up to relabeling, there are two such graph types (see below):

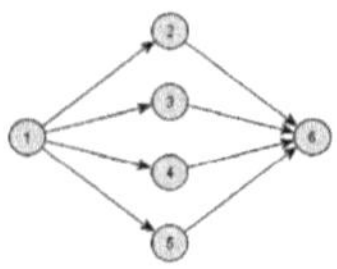

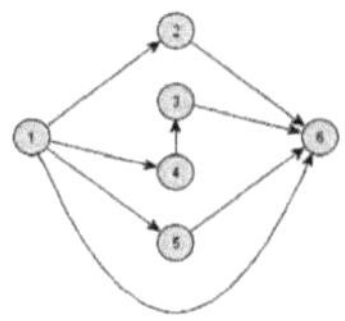

K=25	accept all process avg. $n(\mathbf{c})$	correct process avg. $n(\mathbf{c})$
predict	58.52	47.92
actual	58.32	47.95

The left-hand graph type has mobility $K(K-1)$, and it occurs in only one network configuration. The right-hand graph type has mobility $2K-3$ and multiplicity $K(K-1)$. The table shows the results of numerical simulations for $K = 25$, compared to theoretical predictions. We have used the mobility itself as a marker for the proportion of time spent in each type of configuration; the difference between the two processes is striking, and it will translate into differences in measurements of any observable that differentiates between the two configurations.

4.6 Generation of Random Graphs with Prescribed Degree Correlations via Rewiring Algorithms

Our approach can be extended to accurately target desired degree correlations defined by $W(k, k')$. This can be achieved for undirected graphs by ensuring convergence of the above Markov chain to the following nonuniform measures [8]:

$$p(\mathbf{c}) = \frac{\delta_{\mathbf{k},\mathbf{k}(\mathbf{c})}}{Z} \prod_{i<j} \left[\frac{\langle k \rangle}{N} \frac{W(k_i,k_j)}{p(k_i)p(k_j)} \delta_{c_{ij},1} + \left(1 - \frac{\langle k \rangle}{N} \frac{W(k_i,k_j)}{p(k_i)p(k_j)}\right) \delta_{c_{ij},0} \right] \quad (4.23)$$

with $p(k) = N^{-1} \sum_i \delta_{k_i,k}$, $W(k,k') = (N\langle k \rangle)^{-1} \sum_{ij} c_{ij} \delta_{k_i,k} \delta_{k_j,k'}$, $\langle k \rangle = \sum_k kp(k)$, and

$$W(k) = \sum_{k'} W(k,k') = p(k)k/\langle k \rangle. \quad (4.24)$$

In the language of our process, (4.23) corresponds to the choice

$$H(\mathbf{c}) = -\sum_{i<j} \log \left[\frac{\langle k \rangle}{N} \frac{W(k_i,k_j)}{p(k_i)p(k_j)} \delta_{c_{ij},1} + \left(1 - \frac{\langle k \rangle}{N} \frac{W(k_i,k_j)}{p(k_i)p(k_j)}\right) \delta_{c_{ij},0} \right]. \quad (4.25)$$

Hence, for the candidate edge swaps $\mathbf{c}' \to \mathbf{c} = F_{ijk\ell;\alpha}\mathbf{c}'$, the acceptance probability (4.4) can be used, where

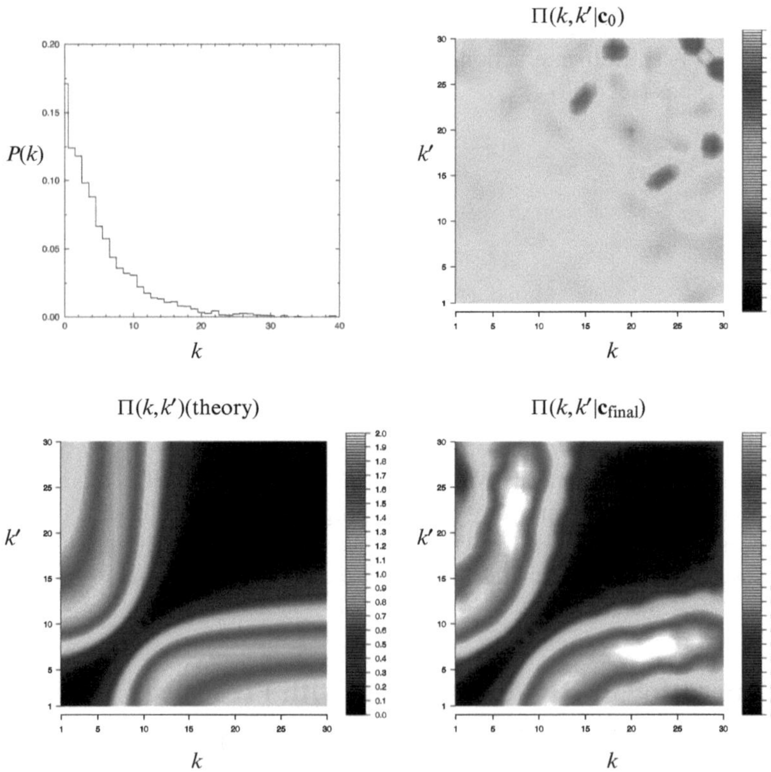

Fig. 4.1 Results of Markovian dynamics tailored to target the non-uniform measure (4.23). *Top left*: degree distribution of the (randomly generated) initial graph $\mathbf{c}_0$, with $N = 4{,}000$ and $\langle k \rangle = 5$. *Top right*: $\Pi(k,k'|\mathbf{c}_0)$ of the initial graph. *Bottom left*: the target relative degree correlations (4.27) chosen in (4.23). *Bottom right*: colour plot of $\Pi(k,k'|\mathbf{c}_{\text{final}})$ in the final graph $\mathbf{c}_{\text{final}}$, measured after 75,000 accepted moves of the Markov chain (4.14)

$$e^{H(\mathbf{c})-H(\mathbf{c}')} = \prod_{(a,b)\in S_{ijk\ell;\alpha}} \left[L_{ab}\delta_{c_{ab},1} + L_{ab}^{-1}\delta_{c_{ab},0}\right] \tag{4.26}$$

with $L_{ab} = N\bar{k}/[\Pi(k_a,k_b)k_ak_b] - 1$ and relative degree correlations $\Pi(k,k') = W(k,k')/W(k)W(k')$.

We show an example of the Markov chain (4.14) targeting (4.23), where relative degree correlations are chosen as

$$\Pi(k,k') = \frac{(k-k')^2}{[\beta_1 - \beta_2 k + \beta_3 k^2][\beta_1 - \beta_2 k' + \beta_3 k'^2]} \tag{4.27}$$

[the parameters β_i follow from (4.24)]. An initial graph $\mathbf{c}_0$ was constructed with a non-Poissonian degree distribution and no degree correlations, i.e., $\Pi(k,k'|\mathbf{c}_0) \approx 1$. After iterating the Markov chain until equilibrium (after 75,000 accepted moves, and after reaching maximal Hamming distance between initial and final configurations), degree correlations are seen to be in very good agreement with their target values; see Fig. 4.1. Extension to directed graphs is achieved by replacing k with $\mathbf{k} = (k^{\text{in}}, k^{\text{out}})$, allowing repetition of site indices in (4.26) and bearing in mind that $\Pi(\mathbf{k}_a,\mathbf{k}_b) \neq \Pi(\mathbf{k}_b,\mathbf{k}_a)$.

4.7 Discussion

In this paper, we have focused on how to generate numerically tailored random graphs with controlled macroscopic structural properties, to serve, e.g., as null models in hypothesis testing. Bias in the generation of random graphs has the potential to invalidate all further statistical analysis performed on the generated networks and has been well documented in the literature; it is known to affect the "stubs" method and the "accept-all" edge-swapping method. Our theory offers a practical and theoretically sound approach to uniformly generating random graphs from ensembles that share certain topological characteristics with a real network, and can therefore serve as a reliable tool for building unbiased null models.

References

1. Bender, E., Canfield, E.: The asymptotic of labelled graphs with given degree sequences. J. Comb. Theory, Ser. A **24**, 296–307 (1978)
2. Chung, F., Lu, L.: The average distances in random graphs with given expected degrees. Proc. Natl. Acad. Sci. USA **99**, 15879–15882 (2002)
3. Coolen, A.C.C., De Martino, A., Annibale, A.: Constrained Markovian dynamics of random graphs. J. Stat. Phys. **136**(6), 1035–1067 (2009)
4. Eggleton, R.B, Holton, D.A.: Simple and multigraphic realizations of degree sequences. In: McAvaney, K.L. (ed.) Combinatorial Mathematics VIII. Lecture Notes in Mathematics, pp. 155–172. Springer, Berlin (1981)

5. Klein-Henning, H., Hartman, A.K.: Bias in generation of random graphs. Phys. Rev. E **85**, 026101 (2012)
6. Molloy, M., Reed, B.: A critical point for random graphs with a given degree sequence. Random Struct. Algor. **6**, 161–180 (1995)
7. Newman, M.E.J., Strogatz, S.H., Watts, D.J.: Random graphs with arbitrary degree distributions and their applications. Phys. Rev. E **64**, 026118 (2001)
8. Perez-Vicente, C.J., Coolen, A.C.C.: Spin models on random graphs with controlled topologies beyond degree constraints. J. Phys. A **41**, 255003 (2008)
9. Rao, A.R., Jana, R., Bandyopadhyay, S.: A Markov chain Monte Carlo method for generating random (0, 1)-matrices with given marginals. Sankhya Ser. A, **58**(2), 225–242 (1996)
10. Roberts, E.S., Coolen, A.C.C.: Unbiased degree preserving randomization of directed binary graphs. `http://arxiv.org/pdf/1112.4677` (2011)
11. Stauffer, A.O., Barbosa, V.C.: A study of the edge switching Markov-chain method for the generation of random graphs. `http://arXiv.org/pdf/0512105` (2005)
12. Taylor, R.: Constrained switchings in graphs. In: McAvaney, K.L. (ed.) Combinatorial Mathematics VIII, Lecture Notes in Mathematics, pp. 314–336. Springer, Berlin (1981)

Chapter 5
A Case Leading to Rationality of the Drift

Gastão Bettencourt

Abstract Let G be an infinite finitely generated group endowed with a measure of probability and a left-invariant metric. Let ∂G be the horofunction compactification. Using a representation of the drift via horofunctions, we establish one situation in which we have rationality of the drift.

5.1 Introduction

Let G be an infinite finitely generated group, endowed with a metric d and a measure of probability μ. With some natural hypothesis on the metric space (G,d), we can attach a boundary to G. Here we will use the horofunction boundary ∂G. If we form products $Z_n = g_o g_1 \cdots g_{n-1}$ of elements of G, the number that describes the way in which these products go to infinity is the drift, defined by

$$\ell(\mu) = \lim_{n \to +\infty} \frac{d(Z_n, e)}{n}.$$

The drift depends on μ, the measure that describes the way we choose the elements to form the product Z_n. We use a fundamental result of Karlsson and Ledrappier that allows us to represent the drift as a sum of horofunctions. Assuming that the boundary ∂G is countable, we obtain a condition for the rationality of the drift.

G. Bettencourt (✉)
UBI – Universidade da Beira Interior, Av. Marquês d' Àvila e Bolama,
6200-001 Covilhã, Portugal
e-mail: gastao@ubi.pt

C. Grácio et al. (eds.), *Nonlinear Maps and their Applications*, Springer Proceedings in Mathematics & Statistics 57, DOI 10.1007/978-1-4614-9161-3_5,

5.2 Background

We recall briefly the standard notions and constructions that we will need. All the background material is given in the references [3–5] and in particular in [2]. Given a proper noncompact metric space (X,d), there are several ways of compactifying it, that is, to put X inside some larger compact space. We explain briefly the so-called Gromov compactification, or horofunction compactification. To obtain this compactification, we begin with the function $\Phi : X \to C(X)$,

$$\Phi(x) = d(x,\cdot) - d(x,x_o),$$

where x_o is a fixed point, and $C(X)$ is the space of continuous functions of X endowed with the topology of uniform convergence on compact sets. The limit of the sequence of continuous functions

$$h(\cdot) = \lim_{x_n \to \infty} d(x_n,\cdot) - d(x_n,x_o)$$

is called a horofunction, and it is a common concept in several branches of geometry. The compactification is the space $\overline{\Phi(X)} = \Phi(X) \cup \partial X$, where $\Phi(X)$ is identified with X, and the elements in ∂X are the horofunctions.

In the present situation, we are interested in applying this construction to the case that the metric space X is a group G. We will assume several hypotheses on G, which will be a locally compact group with a left-invariant proper metric d and a nondegenerate probability measure μ, and finally, we will assume that μ has a finite first moment, i.e.,

$$\int_G d(e,g)d\mu(g) < \infty.$$

We can choose n elements of G and form their product. If we use the law μ to extract independently the elements of the group, then we are in the presence of a so-called random walk on the group. Let $Z_n = g_o g_1 \cdots g_{n-1}$. The drift of the measure μ is defined as

$$\ell(\mu) = \lim_{n \to +\infty} \frac{1}{n} d(e,Z_n).$$

Recall that G acts on the elements of ∂G. If in ∂G we have a measure λ, then the convolution between μ and λ is defined by

$$\mu * \lambda(h) = \int_{g \in G} \lambda(g^{-1}h)d\mu(g).$$

The measure λ is said to be $\mu-$stationary if $\mu * \lambda(h) = \lambda(h)$ for all $h \in \partial G$.

Let now B be a countable set endowed with a probability measure λ. In a classical argument explained in [1], it is proved that if the group G acts on B, the measure μ is nondegenerate, and λ is μ-stationary, then the orbit associated with the maximal point of B is finite. We are now in a position to state a fundamental result of Karlsson and Ledrappier [4].

Theorem 1 (Karlsson–Ledrappier). *Let G be a locally compact group with a left-invariant proper metric d, and let μ be a nondegenerate probability measure on G with finite first moment. Let $\overline{G}$ be the horofunction compactification of (G,d). Then there exists a measure λ on $\overline{G}$ with the following properties:*

- *λ is μ-stationary.*
- *We have the Furstenberg–Khasminskii formula for the linear drift:*

$$\ell(\mu) = \int h(g^{-1})\, d\lambda(h)\, d\mu(g).$$

Moreover, if $\ell(\mu) > 0$, then λ is supported on ∂G.

5.3 Rationality of the Drift

We assume that G is an infinite finitely generated group that satisfies all the natural assumptions on G such that the theorem stated above applies. We restrict ourselves to the case that ∂G, the horofunction boundary, is a countable set. We begin by realizing that the case $\ell(\mu) = 0$ is trivial, so we may assume that $\ell(\mu) > 0$. In this case, we know already that the measure λ is supported on ∂G, a fact that we will use. Using the Karlsson–Ledrappier representation of the drift, we have

$$\begin{aligned} \ell(\mu) &= \int h(g^{-1})\, d\lambda(h)\, d\mu(g) \\ &= \sum_{g \in G} \sum_{h \in \partial G} h(g^{-1}) \lambda(h) \mu(g). \end{aligned}$$

If the measure λ is uniform, the support of λ has to be finite (by an argument of Ballmann), and for some $h_o \in \partial G$, we get

$$\ell(\mu) = \lambda(h_o) \sum_{g_i \in G} \sum_{h_j \in \partial G} h_j(g^{-1}) \mu(g_i).$$

If the measure μ has finite support and takes rational values, that is, $\mu(g) \in \mathbb{Q}$ for all $g \in G$, then this sum is a finite sum of horofunctions evaluated at different points multiplied by rational coefficients.

If the distance considered in the group is the distance generated by the word length, the horofunctions are integer-valued functions, and we arrive at the following result.

Proposition 1. *Let G be a finitely generated group, μ a nondegenerate measure on G with finite first moment, and d the distance generated by the word length. Let ∂G be a countable set.*

If μ has finite support and assumes rational values and if λ is uniform, then the drift is rational.

References

1. Ballmann, W.: Lectures on Spaces of Nonpositive Curvature. DMV. Birkhäuser, Basel (1995)
2. Karlsson, A.: Ergodic theorems for noncommutting random products, Lectures notes from Santiago and Wroclaw, unpublished (2008)
3. Karlsson, A., Ledrappier, F.: Linear drift and Poisson boundary for random walks. Pure Appl. Math. Q., Pt 1 **3**(4), 1027–1036 (2007)
4. Karlsson, A., Ledrappier, F.: Noncommutative ergodic theorems. In: Farb, B., Fisher, D. (eds.) Geometry, Rigidity, and Group Actions. Chicago Lectures in Mathematics Series, vol. 75. The University of Chicago Press, Chicago (2011)
5. Karlsson, A., Ledrappier, F.: On laws of large numbers for random walks. Ann. Prob. **34**(5), 1693–1706 (2006)

Chapter 6
Remarks on the Nonwandering Set of Skew Products with a Closed Set of Periodic Points of the Quotient Map

Lyudmila S. Efremova

Abstract Exact formulas are given for the description of the nonwandering set of a continuous skew product of interval maps with a closed set of periodic points in the base. Special multifunctions, such as the Ω-function and suitable functions for the Ω-function, are used. A corollary is given for the nonwandering set of the C^1-smooth skew product of interval maps with a closed set of periodic points.

6.1 Introduction

Dynamical systems of skew product class are arising in investigations of mathematical models of totally developed turbulence [4], signal processing [9], quasicrystal physics [5], and others. In this paper, the structure of the nonwandering set of skew products of maps of an interval with a closed set of periodic points of the quotient map is investigated.

Let $I = I_1 \times I_2$ be a closed rectangle in the plane (I_1, I_2 are closed intervals). We consider a *skew product of interval maps* $F : I \to I$, i.e., a dynamical system

$$F(x, y) = (f(x), g_x(y)), \text{ where } g_x(y) = g(x, y), \ (x; y) \in I. \tag{6.1}$$

Here $f : I_1 \to I_1$ is said to be the *quotient map* of the dynamical system (6.1), and the map $g_x : I_2 \to I_2$ for $x \in I_1$ is said to be the *map acting on the fiber* over the point x.

As follows from (6.1), the equality

$$F^n(x, y) = (f^n(x), g_{x,n}(y)), \text{ where } g_{x,n} = g_{f^{n-1}(x)} \circ \cdots \circ g_{f(x)} \circ g_x, \tag{6.2}$$

L.S. Efremova (✉)
Department of Differential Equations and Mathematical Analysis, Faculty of Mathematics and Mechanics, University of Nizhny Novgorod, Gagarin Ave.,
23a, build. 6., 603950 Nizhny Novgorod, Russia
e-mail: lefunn@gmail.com

C. Grácio et al. (eds.), *Nonlinear Maps and their Applications*, Springer Proceedings in Mathematics & Statistics 57, DOI 10.1007/978-1-4614-9161-3_6,

holds for every point $(x; y) \in I$ and natural number n. We use the notation $\tilde{g}_x$ for a map $g_{x,n}$ if x is a periodic point of the quotient map f ($x \in Per(f)$), and n is its (least) period. Denote by $T^0(I)$ ($T^1(I)$) the space of all continuous (C^1-smooth) skew products of maps of an interval with C^0-norm (C^1-norm).

The nonwandering set (see, e.g., [22]) plays an important role in the investigation of the properties of an arbitrary dynamical system with a compact phase space.

Definition 1. A point $z^0(x^0; y^0) \in I$ is called a *nonwandering point* for a map $F \in T^0(I)$ if for every neighborhood $U(z^0)$ of the point z^0 in I, there exists a natural number $n = n(U(z^0))$ such that $U(z^0) \bigcap F^n(U(z^0)) \neq \emptyset$. The set of all nonwandering points of the dynamical system (6.1) is called the *nonwandering set* (the notation $\Omega(F)$ is usually used for the nonwandering set). A point $z' \in I \setminus \Omega(F)$ is called a *wandering point* of F.

The structure of the nonwandering set of skew products of maps of an interval with a closed set of periodic points in the base is investigated in [10, 12], and the structure of the nonwandering set of skew products of maps of an interval with a closed set of periodic points is investigated independently in [24] and, to a certain degree, in [3].

Although the papers [10, 12] contain examples of the skew products of maps of an interval from the above class that demonstrate the influence of Ω-explosions in maps acting on the fibers over periodic points of their quotient maps on the structure of the nonwandering sets, nevertheless the author did not notice the possibility of the influence of maps in the fibers over wandering points of the quotient map on the structure of the nonwandering set of a considering skew product.

In the recent paper [19], an example of the skew product of maps of an interval with a closed set of periodic points in the base is given, illustrating the influence of maps acting on the fibers over wandering points of the quotient map on the structure of the nonwandering set of a skew product. In [19], it is indicated also that every point of the phase space of a skew product that is not a chain-recurrent point is a wandering point. This statement gives an upper estimate on the nonwandering set of a skew product, but it does not explain the role of the fiber maps in the formation of the nonwandering set. Moreover, we will give an example showing that this estimate can be degenerate: a skew product with the closed set of periodic points will be constructed such that every point of its phase space is chain recurrent, but the nonwandering set is the polygonal line consisting of three components.

In this paper, we give explicit formulas for the nonwandering set of a skew product of maps of an interval with a closed set of periodic points in the base. To accomplish this, we use the concepts of the Ω-function and auxiliary (suitable) functions for the Ω-function [11, 14].

The paper is organized as follows. In Sect. 6.2, we formulate the main definitions of the Ω-function and auxiliary (suitable) functions for the Ω-function to be used and give an example to illustrate these definitions. In Sect. 6.3, we introduce the concept of weakly nonwandering points with respect to the family of fiber maps, and formulate and prove the main theorem containing explicit formulas for the nonwandering set of a skew product of interval maps with a closed set of periodic points in the base. In Sect. 6.4, we investigate the influence of the differentiability properties on the structure of the nonwandering set of a skew product of maps of an interval with a closed set of periodic points.

6.2 Preliminaries

We begin by giving definitions of the principal multifunctions that will play an important role in the description of the structure of the nonwandering set of a skew product of maps of an interval.

Let 2^{I_2} denote the topological space of all closed subsets of the closed interval I_2 with the exponential topology [25].

Definition 2 (See [11]). The *Ω-function of a map $F \in T^0(I)$* is the function $\zeta^F : \Omega(f) \to 2^{I_2}$ satisfying the equality

$$\zeta^F(x) = (\Omega(F))(x) \quad \text{for every } x \in \Omega(f),$$

where $\Omega(f)$ is the nonwandering set of the quotient map f and

$$(\Omega(F))(x) = \{y \in I_2 : (x, y) \in \Omega(F)\}$$

is the slice of $\Omega(F)$ by the vertical fiber over a point $x \in \Omega(f)$.

By the equality $\Omega(f) = pr_1(\Omega(F))$ (see, e.g., [10, 23]), where $pr_1 : I \to I_1$ is the natural projection, the Ω-function of a map $F \in T^0(I)$ is well defined. We also note that the graph of the Ω-function in the phase space I coincides with the nonwandering set of the skew product F.

We would like to remark that if the nonwandering set $\Omega(F)$ of a skew product F is a global attractor, and the slices $(\Omega(F))(x)$ for every $x \in \Omega(f)$ are closed intervals, then the Ω-function coincides, up to the phase space and a considered dynamical system, with the multifunction introduced by Stark [31] and discussed in [17, 20, 21, 32].

Because the nonwandering set $\Omega(F)$ is closed, the interval I_2 is compact, and the interval I_1 is Hausdorff, it follows that the Ω-function of the map $F \in T^0(I)$ is upper semicontinuous from above [25]. This property of the Ω-function contains information about the topological structure of the nonwandering set of the skew product. The set of discontinuity points of the Ω-function, if nonempty, is the boundary subset of the compact set $\Omega(f)$ of the first Baire category, and the complementary set of continuity points of the Ω-function is the set of the second Baire category [25], which is nonempty for every map $F \in T^0(I)$ and everywhere dense in $\Omega(f)$ [25].

Consider a skew product $F \in T^0(I)$ such that the equality $\Omega(f) = \Omega(f^n)$ holds for every $n > 1$ (skew products of maps of an interval with a closed set of periodic points in the base satisfy this condition). We then have well-defined *auxiliary multifunctions* $\eta_n : \Omega(f) \to 2^{I_2}$ such that the equality $\eta_n(x) = \Omega(g_{x,n})$ holds for each

$x \in \Omega(f)$, where $\Omega(g_{x,n})$ is the nonwandering set of the map $g_{x,n}$, $n \geq 1$ [11], as well as *suitable multifunctions* $\overline{\eta}_n : \Omega(f) \to 2^{I_2}$ for the Ω-function of the map $F \in T^0(I)$ such that $\overline{\eta}_n(x) = (\overline{\eta}_n)(x)$ for every $x \in \Omega(f)$, where $\overline{\eta}_n$ is the closure of the graph of the multifunction η_n and $(\overline{\eta}_n)(x)$ is the slice of the graph of the multifunction $\overline{\eta}_n$ by the fiber over a point x [14].

Using auxiliary multifunctions η_n and direct products $F_{n,1}(x,y) = (f^n(x), id(y))$ (here $F_{n,1} : I \to I$, $id(y)$ is the identity map on the interval I_2), we introduce multifunctions $\eta_{n,1} : \Omega(f) \to 2^{I_2}$ $(n \geq 1)$, setting $\eta_{n,1}(x) = (F_{n,1}(\eta_n))(x)$ for every $x \in \Omega(f)$, where $(F_{n,1}(\eta_n))(x)$ is the slice of the set $F_{n,1}(\eta_n)$ by the fiber over a point $x \in \Omega(f)$. Then the following equality is valid:

$$\eta_{n,1}(x) = \bigcup_{\overline{x} \in \{f^{-n}(x)\}} \eta_n(\overline{x}), \tag{6.3}$$

where $\{f^{-n}(x)\}$ is the nth complete preimage of a point x under $f_{|\Omega(f)}$.

Example 1. Let $F(x,y) = (f(x), x)$ be the skew product of the maps of the unit interval with $f(x) = 4x(1-x)$. Then we have that the Jacobian of this map is zero,

$$F([0,1]^2) = \{(x,y) \in [0,1]^2 : x = 4y(1-y)\} = \Omega(F);$$

for all $x \in [0,1] = \Omega(f)$, and for $n \geq 1$, the following equalities hold:

$$\zeta^F(x) = \left\{ \frac{1-\sqrt{1-x}}{2}; \ \frac{1+\sqrt{1-x}}{2} \right\};$$

$$\eta_n(x) = \{f^{n-1}(x)\}; \quad \eta_{n,1}(x) = \bigcup_{\overline{x} \in \{f^{-n}(x)\}} \{f^{n-1}(\overline{x})\} = \{f^{-1}(x)\}.$$

Then the equality $\zeta^F = \eta_{n,1}$ holds for the graphs (in the phase space I) of the multifunctions ζ^F and $\eta_{n,1}$ for each $n \geq 1$. Note that in this case, $\zeta^F = \underset{n \to +\infty}{Lim}\, \eta_{n,1}$, where $Lim(\cdot)_n$ is the topological limit of a sequence of sets [25].

We have that if $\Omega(f) \neq I_1$, the multifunctions η_n and $\eta_{n,1}$ admit natural extensions η_n^{ex} on the interval I_1 and $\eta_{n,1}^{ex}$ on the interval $f^n(I_1)$ $(n \geq 1)$ respectively. So $\eta_n^{ex}(x) = \Omega(g_{x,n})$ for each $x \in I_1$. Multifunctions $\eta_{n,1}^{ex}$ are defined by the equalities $\eta_{n,1}^{ex}(x) = (F_{n,1}(\eta_n^{ex}))(x)$ for all $x \in f^n(I_1)$.

Consider a skew product $F \in T^0(I)$ such that its quotient map f has a closed set of periodic points $Per(f)$. Then f satisfies the following properties:

1. $\tau(f) = \{1, 2, 2^2, \ldots, 2^\nu\}$, where $\tau(f)$ is the set of the (least) periods of f-periodic points, $0 \leq \nu \leq +\infty$ [29].
2. For every point $x^0 \in Per(f)$, there exists a neighborhood $U_1(x^0) \subset I_1$ satisfying

$$U_1(x^0) \bigcap f^n(U_1(x^0)) \neq \emptyset$$

iff n is a multiple of the (least) period $n(x^0) = 2^{i_0}$ of the point x^0 [29].

3. $Per(f) = \Omega(f)$ [29].
4. The ω-limit set $\omega_f(x)$ of every point $x \in I_1$ is a periodic orbit [29].
5. If a sequence of f-periodic points converges to a point x^0, then the sequence of the trajectories of these points converges to the trajectory of x^0 [16].
   We need also the characteristic property of nonwandering points of a continuous map of an interval:
6. For every map $g \in C^0(I_2)$ the following holds: $y \in \Omega(g)$ iff for every neighborhood $U_2(y)$ of the point y in I_2, there are $y' \in U_2(y)$ and $m' \geq 1$ such that $g^{m'}(y') = y$ [28].

Denote by $Per(f, l_n)$ the set of f-periodic points such that the (least) period of each of them divides $l_n = 2^n$, $n \geq 0$ (i.e., the (least) periods of these points form the set $\{1, 2, 2^2, \ldots, 2^n\}$). The following hold for every $n \geq 0$:[1]

$$\begin{aligned} &\eta'_{l_n|Per(f,l_n)} = \bigcup_{i=0}^{n} \eta_{l_i|Per(f,l_i)}, \quad \eta'_{l_n,1|Per(f,l_n)} = \bigcup_{i=0}^{n} \eta_{l_i,1|Per(f,l_i)}, \\ &\eta^{ex'}{}_{l_n,1} = \bigcup_{i=0}^{n} \eta^{ex}_{l_i,1}. \end{aligned} \tag{6.4}$$

In addition, the last equality from (6.4) is valid for all $x \in \bigcap_{i=0}^{n} f^{l_i}(I_1)$.

6.3 Weakly Nonwandering Points with Respect to the Family of Fibers Maps

Next, we prove the following result.

Theorem 1. *Let a map $F \in T^0(I)$ have the quotient map f with a closed set $Per(f)$.*

Then the topological limits $\underset{n\to+\infty}{Lim}\, \eta'_{l_n,1|Per(f,l_n)}$ and $\underset{n\to+\infty}{Lim}\, \eta'_{l_n|Per(f,l_n)}$ exist and satisfy

$$\zeta^{F_{|Per(f)\times I_2}} = \underset{n\to+\infty}{Lim}\, \eta'_{l_n,1|Per(f,l_n)} = \underset{n\to+\infty}{Lim}\, \eta'_{l_n|Per(f,l_n)} = \overline{\bigcup_{x\in Per(f)} \{x\} \times \Omega(\tilde{g}_x)}, \tag{6.5}$$

where $\eta'_{l_n,1|Per(f,l_n)}$, $\eta'_{l_n|Per(f,l_n)}$, $\zeta^{F_{|Per(f)\times I_2}}$ are the graphs of the corresponding multifunctions in I.

Moreover, if x is a periodic point of f with (least) period $l(x)$ that is the limit point for the aperiodic points of f, then we have

$$(\zeta^F)(x) = \underset{n\to+\infty}{Ls}\, \eta^{ex'}_{l(x)n,1|U_{1,\varepsilon_{l(x)n}}}(x), \tag{6.6}$$

[1] By introducing the multifunctions η'_{l_n} and $\eta'_{l_n,1}$, one avoids complications about the possible breakdown of the equality $\Omega(\tilde{g}_x^{2^i}) = \Omega(\tilde{g}_x^{2^{i-1}})$ (see [8]).

where $\underset{n\to+\infty}{Ls}(\cdot)_n$ *is the topological upper limit of a sequence of sets [25] and* $U_{1,\varepsilon_{l(x)n}}(x)$ *is an* $\varepsilon_{l(x)n}$*-neighborhood of the point* $x \in Per(f)$ *such that* $\lim_{n\to+\infty} \varepsilon_{l(x)n} = 0$.[2]

We need an auxiliary claim, which is a direct corollary of property 2 from Sect. 6.2 and Definition 1; see [10, 12].

Lemma 1. *Let the set* $Per(f)$ *of the quotient map* f *of a skew product* $F \in T^0(I)$ *be closed. Then for an* f*-periodic point* x^0 *with period* $n(x^0) = 2^{i_0}$ $(i_0 \geq 1)$*, the following claims are equivalent:*

(a) $(x^0, y^0) \in \Omega(F)$.
(b) $(x^0, y^0) \in \Omega(F^{n(x^0)})$.

We will prove the first part of Theorem 1, that is, we shall give the proof of the equality (6.5), in the two steps indicated in Lemmas 2 and 3. We use not only the properties of continuous maps of an interval with a closed set of periodic points (see Sect. 6.2) but also the Ω-function and multifunctions η'_{l_n}, $\eta'_{l_n,1}$. By property 3, the multifunctions η'_{l_n} and $\eta'_{l_n,1}$ are defined on the set $Per(f)$.

Lemma 2. *Let the set* $Per(f)$ *of the quotient map* f *of a skew product* $F \in T^0(I)$ *be closed. Then there exist the topological limits* $\underset{n\to+\infty}{Lim}\, \eta'_{l_n,1\,|Per(f,l_n)}$ *and* $\underset{n\to+\infty}{Lim}\, \eta'_{l_n\,|Per(f,l_n)}$*. Moreover,*

$$\underset{n\to+\infty}{Lim}\, \eta'_{l_n,1\,|Per(f,l_n)} = \underset{n\to+\infty}{Lim}\, \eta'_{l_n\,|Per(f,l_n)} = \overline{\bigcup_{x\in Per(f)} \{x\} \times \Omega(\tilde{g}_x)}, \tag{6.7}$$

where $\{\eta'_{l_n,1\,|Per(f,l_n)}\}_{n\geq 1}$ $(\{\eta'_{l_n\,|Per(f,l_n)}\}_{n\geq 1}$ *) is the sequence of graphs of the multifunctions* $\eta'_{l_n,1\,|Per(f,l_n)}$ $(\eta'_{l_n\,|Per(f,l_n)})$ *in* I.

Proof. By the first formula (6.4), the inclusions

$$\eta'_{l_n\,|Per(f,l_n)} \subseteq \eta'_{l_{n+1}\,|Per(f,l_{n+1})} \tag{6.8}$$

hold for every $n \geq 0$.

The inclusions (6.8) imply the existence of the topological limit $\underset{n\to+\infty}{Lim}\, \eta'_{l_n\,|Per(f,l_n)}$ (see [25]) and correctness of the equality

$$\underset{n\to+\infty}{Lim}\, \eta'_{l_n\,|Per(f,l_n)} = \overline{\bigcup_{x\in Per(f)} \{x\} \times \Omega(\tilde{g}_x)}. \tag{6.9}$$

---

[2] This theorem in the above form is announced in [15]. As the equality (6.5) shows, results concerning the structure of the nonwandering set from [10, 12] are correct for the restriction of the map F on the set $Per(f) \times I_2$.

In the case of a bounded set $\tau(f)$ (this means that $\nu < +\infty$), we have

$$\eta'_{l_n|Per(f,l_n)} = \eta'_{l_\nu|Per(f,l_\nu)} \tag{6.10}$$

for all $n \geq \nu$.

By (6.3) and (6.4), the equalities

$$\eta_{l_{n,1}|Per(f,l_n)} = \eta_{l_n|Per(f,l_n)}, \quad \eta'_{l_{n,1}|Per(f,l_n)} = \eta'_{l_n|Per(f,l_n)} \tag{6.11}$$

are valid for all $n \geq 0$. Then from (6.8), (6.9), and (6.11), we have

$$\underset{n\to+\infty}{Lim}\, \eta'_{l_{n,1}|Per(f,l_n)} = \underset{n\to+\infty}{Lim}\, \eta'_{l_n|Per(f,l_n)} = \overline{\bigcup_{x\in Per(f)} \{x\} \times \Omega(\tilde{g}_x)}. \tag{6.12}$$

If the set $\tau(f)$ is bounded, then using formulas (6.10) and (6.11), we get

$$\underset{n\to+\infty}{Lim}\, \eta'_{l_{n,1}|Per(f,l_n)} = \overline{\eta'}_{l_\nu|Per(f,l_\nu)} = \overline{\eta'}_{l_\nu} = \overline{\bigcup_{x\in Per(f)} \{x\} \times \Omega(\tilde{g}_x)}, \tag{6.13}$$

and Lemma 2 is proved. □

Lemma 3. *If the conditions of Lemma 2 are fulfilled, then the equality*

$$\zeta^{F_{|Per(f)\times I_2}} = \underset{n\to+\infty}{Lim}\, \eta'_{l_{n,1}|Per(f,l_n)} \tag{6.14}$$

is valid.

Proof. By Lemma 2, we know that the set $\tilde{\Omega}(F_{|Per(f)\times I_2}) = \underset{n\to+\infty}{Lim}\, \eta'_{l_{n,1}|Per(f,l_n)}$ is well defined. Therefore, we have only to prove that

$$\tilde{\Omega}(F_{|Per(f)\times I_2}) = \Omega(F_{|Per(f)\times I_2}) = \zeta^{F_{|Per(f)\times I_2}}. \tag{6.15}$$

However, since the inclusion $\tilde{\Omega}(F_{|Per(f)\times I_2}) \subseteq \Omega(F_{|Per(f)\times I_2})$ follows from (6.12), we have only to prove the inverse inclusion $\Omega(F_{|Per(f)\times I_2}) \subseteq \tilde{\Omega}(F_{|Per(f)\times I_2})$.

To this end, we use the set

$$\pi_\infty = \bigcap_{n=1}^{+\infty} F^n(Per(F) \times I_2).$$

Since every point $x \in Per(f)$ has a unique preimage in $Per(f)$ and I_2 is connected and compact, it follows that the slice $(\pi_\infty)(x)$ is a closed interval (possibly degenerate) for every $x \in Per(f)$ [25]. Moreover, $F : \pi_\infty \to \pi_\infty$ is a surjective map, and $\Omega(F_{|Per(f)\times I_2}) \subseteq \pi_\infty$. Therefore the equality (6.15) holds if $\tilde{\Omega}(F_{|Per(f)\times I_2}) = \pi_\infty$.

Let us suppose that $\tilde{\Omega}(F_{|Per(f)\times I_2}) \neq \pi_\infty$. We will show that every point from $\pi_\infty \setminus \tilde{\Omega}(F_{|Per(f)\times I_2})$ wanders in π_∞.

We use the multifunction $\tilde{\zeta} : Per(f) \to 2^{I_2}$, setting $\tilde{\zeta}(x) = (\tilde{\Omega}(F_{|Per(f)\times I_2}))(x)$ for every $x \in Per(f)$. By analogy with the Ω-function, the multifunction $\tilde{\zeta}$ is upper semicontinuous, and its graph is a closed set in I. Then by (6.12), for every point

$(x;y) \in \pi_\infty \setminus \tilde{\Omega}(F_{|Per(f)\times I_2})$, there exists the complementary interval $(a_x(y),\, b_x(y))$ for the slice $(\tilde{\Omega}(F_{|Per(f)\times I_2}))(x)$ in the closed interval $(\pi_\infty)(x)$ containing the point y and satisfying that for every sequence $\{(x_i;y_i)\}_{i\geq 1} \subset \pi_\infty \setminus \tilde{\Omega}(F_{|Per(f)\times I_2})$ converging to $(x;y)$, the inclusion

$$(a_x(y),\, b_x(y)) \subset (a_{x_i}(y_i),\, b_{x_i}(y_i)) \tag{6.16}$$

is valid. Note also that $(a_x(y),\, b_x(y)) \bigcap \Omega(\tilde{g}_x) = \emptyset$.

Since $F_{|\pi_\infty}$ is a surjective map, the complete preimage of $\{x\} \times (a_x(y),\, b_x(y))$ $(F_{|\pi_\infty})^{-n}(\{x\} \times (a_x(y),\, b_x(y)))$ is well defined for every $n \geq 1$ and $x \in Per(f)$. Then we have

$$(F_{|\pi_\infty})^{-n}(\{x\} \times (a_x(y),\, b_x(y))) \bigcap (\{x^{-n}\} \times \Omega(\tilde{g}_{x^{-n}})) = \emptyset, \tag{6.17}$$

where $x^{-n} = (f_{|Per(f)})^{-n}(x)$.

Fix a point $(x^0;y^0) \in \pi_\infty \setminus \tilde{\Omega}(F_{|Per(f)\times I_2})$. Let $\{(x_i;y_i)\}_{i\geq 1} \subset \pi_\infty \setminus \tilde{\Omega}(F_{|Per(f)\times I_2})$ be an arbitrary sequence converging to $(x^0;y^0)$.

We first consider the case

$$(\{x^0\} \times (a_{x^0}(y^0),\, b_{x^0}(y^0))) \bigcap (F_{|\pi_\infty})^{-n}(\{x^0\} \times (a_{x^0}(y^0),\, b_{x^0}(y^0))) = \emptyset \tag{6.18}$$

for $n \geq 1$. In fact, since $Per(f)$ is a closed set, $\lim\limits_{i\to+\infty} x_i = x^0$, it follows that the f-trajectories of periodic points x_i converge to the f-trajectory of the periodic point x^0 (see property 5, Sect. 6.2). This property together with the continuity of F implies the closedness of the set $(F_{|\pi_\infty})^{-n}(\{(x^0;y^0), \{(x_i;y_i)\}_{i\geq 1}\})$ for all $n \geq 1$. Denote by $L_{(x^0;y^0)}$ the set of all limit points of

$$\bigcup_{n=1}^{+\infty} (F_{|\pi_\infty})^{-n}(\{(x^0;y^0), \{(x_i;y_i)\}_{i\geq 1}\}).$$

Then by the above, $L_{(x^0;y^0)} \subset Orb_f(x^0) \times I_2$ ($Orb_f(x^0) = \{x^0, f(x^0), \ldots f^{l(x^0)-1}\}$ is an f-periodic orbit of x^0), and by (6.18), the following inclusion holds:

$$L_{(x^0;y^0)} \bigcap (\{x^0\} \times I_2) \subset (\pi_\infty)(x^0) \setminus (a_{x^0}(y^0),\, b_{x^0}(y^0)).$$

Let

$$\varepsilon_0 = \frac{1}{2} \min\{y^0 - a_{x^0}(y^0), b_{x^0}(y^0) - y^0\}$$

if $l(x^0) = 1$, and

$$\varepsilon_0 = \frac{1}{2} \min\{y^0 - a_{x^0}(y^0), b_{x^0}(y^0) - y^0, \min_{\substack{i\neq j \\ 0\leq i,\, j\leq l(x^0)-1}} \{|f^i(x^0) - f^j(x^0)|\}\},$$

if $l(x^0) \geq 2$. Then we have

$$U_{\varepsilon_0}((x^0; y^0)) \bigcap (F_{|\pi_\infty})^{-n}(U_{\varepsilon_0}((x^0; y^0))) = \emptyset$$

for every $n \geq n_0$ for some $n_0 \geq 1$, where $U_{\varepsilon_0}((x^0; y^0))$ is an ε_0-neighborhood of the point $(x^0; y^0)$ in π_∞. Thus, the point $(x^0; y^0)$ wanders in π_∞.

Consider the alternative case. Then by (6.17) and (6.18), for some n divisible by $l(x^0)$ there is a connected component of the set $(F_{|\pi_\infty})^{-n}(\{x^0\} \times (a_{x^0}(y^0), b_{x^0}(y^0)))$ belonging to the interval $\{x^0\} \times (a_{x^0}(y^0), b_{x^0}(y^0))$. Hence, at least one of the points $a_{x^0}(y^0)$ and $b_{x^0}(y^0)$ is a periodic point of the map $\tilde{g}_{x^0}$. Suppose that $a_{x^0}(y^0)$ is a periodic point of $\tilde{g}_{x^0}$ with least period $m = m(y^0)$ (here $n = l(x^0)m(y^0)$), and $\tilde{g}^m_{x^0}(y) > y$ for every $y \in (a_{x^0}(y^0), b_{x^0}(y^0))$. By the continuity of F, there exists a neighborhood $U_1(x^0)$ of the point x^0 in $Per(f)$ consisting of the entire set of f-periodic orbits (see property 5, Sect. 6.2) satisfying that for every $x \in U_1(x^0)$, there exists a connected component $U_{2,x}(y) \subset \{x^{-n}\} \times I_2$ of the complete preimage $(F_{|\pi_\infty})^{-n}(\{x\} \times (a_x(y), b_x(y))$ of the interval $\{x\} \times (a_x(y), b_x(y))$ such that

$$\bigcup_{x \in U_1(x^0)} U_{2,x}(y) \subset \bigcup_{x \in U_1(x^0)} \{x\} \times (a_x(y), b_x(y)). \tag{6.19}$$

By (6.19), for every $x \in U_1(x^0)$ and $n(x) = l(x)m(y_0)$ [here n divides $n(x)$], there is a connected component of the complete preimage $(F_{|\pi_\infty})^{-n(x)}(\{x\} \times (a_x(y), b_x(y))$ belonging to the interval $\{x\} \times (a_x(y), b_x(y))$. This means that at least one of the points $a_x(y)$ and $b_x(y)$ is a periodic point of the map $\tilde{g}_x$ with least period $m = m(y_0)$.

Since $\tilde{g}^m_{x^0}(y) = g_{x^0,n}(y) > y$ for every $y \in (a_{x^0}(y^0), b_{x^0}(y^0))$, it follows that by the inclusions (6.16) and (6.19), there exists a neighborhood of the point x^0 in $Per(f)$ [we assume that it is $U_1(x^0)$] such that $g_{x,n}(y') > y'$ for $x \in U_1(x^0)$ and $y' \in (a_x(y), b_x(y))$. The last inequality implies $g_{x,n(x)}(y') > y'$ for every $x \in U_1(x^0)$ and $y' \in (a_x(y), b_x(y))$. Using the inclusion (6.19), we guarantee that there is a neighborhood of the point x^0 in $Per(f)$ [as above, we assume that it is $U_1(x^0)$] satisfying

$$\tilde{g}^m(a_x(y)) = g_{x,n(x)}(a_x(y)) = a_x(y) \quad \text{for any } x \in U_1(x^0). \tag{6.20}$$

Let $U_1'(x^0)$ be a neighborhood of the point x^0 such that $\overline{U}_1'(x^0) \subset U_1(x^0)$. Define the closed set $O_f^* = \bigcup_{x \in \overline{U}_1'(x^0)} Orb_f(x)$ (see property 5, Sect. 6.2). From the results above, there exists a unique point

$$b^s = \sup_{x \in O_f^*} \{y \in (a_{x^{-n}}(y^{-n}); y^0) : g_{x^{-n},n}(y^{-n}) = y^0\}, \text{ and } b^s < y^0.$$

Using the uniform continuity of the map $g_{x,n|\pi_\infty}$, for $\varepsilon = (y^0 - b_s)/3$ ($\varepsilon > 0$) we find a positive number $\delta < \min\{\varepsilon, l(U_1'(x^0))\}$ (here $l(\cdot)$ is the length of an interval) such that

$$U_\delta((x^0; y^0)) \bigcap (F_{|\pi_\infty})^{-jn}(U_\delta((x^0; y^0))) = \emptyset \text{ for every } j \geq 1,$$

where $U_\delta((x^0;y^0))$ is a δ-neighborhood of the point $(x^0;y^0)$ in π_∞. Therefore, $(x^0;y^0) \notin \Omega(F^n_{|\pi_\infty})$.

Finally, there is a neighborhood $U_1'(x^0)$ of the point $x^0 \in Per(f)$ in I_1 such that

$$U_1'(x^0) \bigcap f^p(U_1'(x^0)) \neq \emptyset$$

iff $l(x^0)$ divides p (see property 2 in the list of properties of continuous maps of an interval with a closed set of periodic points from Sect. 6.2).

Set $U_1'(x^0) = (\alpha, \beta)$, where $\delta' = \min\{\delta, x^0 - \alpha, \beta - x^0\}$. Then using (6.20), we obtain the following equality for the neighborhood $U_{\delta'}((x^0;y^0))$ of the point $(x^0;y^0)$ in π_∞:

$$U_{\delta'}((x^0;y^0)) \bigcap (F_{|\pi_\infty})^{-j}(U_{\delta'}((x^0;y^0))) = \emptyset \text{ for any } j \geq 1.$$

Thus $(x^0;y^0) \notin \Omega(F_{|\pi_\infty})$, and every point of the set $\pi_\infty \setminus \tilde{\Omega}(F_{|Per(f)\times I_2})$ wanders in π_∞.

Since $pr_1(\pi_\infty) = Per(f)$ and the set $\pi_\infty \setminus \tilde{\Omega}(F_{|Per(f)\times I_2})$ is open in π_∞, it follows that every point from $\pi_\infty \setminus \tilde{\Omega}(F_{|Per(f)\times I_2})$ wanders in $Per(f) \times I_2$.

If $\pi_\infty \neq Per(f) \times I_2$, then by the inclusion $\Omega(F_{|Per(f)\times I_2}) \subseteq \pi_\infty$, an arbitrary point of the set $(Per(f) \times I_2) \setminus \pi_\infty$ wanders in $Per(f) \times I_2$. This completes the proof of (6.5) and Lemma 3. □

Note that suitable multifunctions $\overline{\eta}_n$ for the Ω-function ($n \geq 1$) have a real dynamical sense. In fact, if we introduce the auxiliary continuous skew products

$$F_n(x,y) = (f^*(x), g_{x,n}(y)) : I \to I,$$

where $f^*(x) = x$ for $x \in \Omega(f)$, and $f^*(x) > x$ for $x \notin \Omega(f)$, then it follows from (6.13) (see the proof of Lemma 2) that

$$\overline{\eta}_n = \Omega(F_{n|\Omega(f)\times I_2}) = \zeta^{F_{n|\Omega(f)\times I_2}}$$

for every $n \geq 1$ (here $\overline{\eta}_n$ is the graph of the function $\overline{\eta}_n$ in I).

Weakly nonwandering points play an important role in the structural questions of dynamical systems theory (see, e.g., [2]). As the following results show, it is useful also to introduce the concept of weakly nonwandering points with respect to the family of fiber maps of a dynamical system of a skew product class.

Definition 3. Let a map $F \in T^r(I)$ ($r = 0, 1$) have a quotient map with a closed set $Per(f)$, and $(x;y) \in Per(f) \times I_2$. A point $y \in I_2$ is said to be a *weakly nonwandering point with respect to the family of fiber maps* of a skew product $F \in T^r(I)$ if

$$(x;y) \in \underset{n\to+\infty}{Ls}\ \eta^{ex'}_{l(x)n,1\,|U_{1,\varepsilon_{l(x)n}}(x)}$$

for an arbitrary sequence $\{\varepsilon_{l(x)n}\}_{n\geq 1}$ of positive numbers such that $\lim_{n\to+\infty} \varepsilon_{l(x)n} = 0$ (see the notations of Theorem 1).

Next, we give examples of C^1-smooth skew products of maps of an interval with a closed set of periodic points in the base containing weakly nonwandering points of the family of fiber maps such that

(i) $y \notin \Omega(\tilde{g}_x)$, but $(x; y) \in \zeta^{F_{|Per(f)\times I_2}} = \overline{\bigcup_{x' \in Per(f)} \{x'\} \times \Omega(\tilde{g}_{x'})}$ (Example 2);

(ii) $y \notin (\zeta^{F_{|Per(f)\times I_2}})(x)$, but $(x; y) \in \underset{n\to+\infty}{Ls}\, \eta^{ex'}_{l(x)n,1\,|U_{1,\varepsilon_{l(x)n}}(x)\setminus Per(f)}$ for every sequence of positive numbers $\{\varepsilon_{l(x)n}\}_{n\geq 1}$ satisfying $\lim_{n\to+\infty} \varepsilon_{l(x)n} = 0$ (Example 3).

Let us consider the following C^1-smooth function:

$$\psi(y) = \begin{cases} -1, & \text{if } y \in [-1, -\frac{1}{2}); \\ 4y(1-|y|), & \text{if } y \in [-\frac{1}{2}, \frac{1}{2}); \\ 12(\frac{1}{4}-y)(y-\frac{3}{4})+\frac{1}{4}, & \text{if } y \in [\frac{1}{2}, \frac{3}{4}); \\ 128(1-y)^4-4(1-y)^2, & \text{if } y \in [\frac{3}{4}, 1]. \end{cases}$$

Note that $\psi(1-\sqrt{\frac{1}{32}}) = 0$. Set $y_0 = 1-\sqrt{\frac{1}{32}}$. Then $\psi^n(y) \in (-1, 0]$ for $y \in (y_0, 1]$ and $n \geq 1$. Hence, all points of the set $\bigcup_{n=0}^{+\infty} \psi^{-n}((y_0, 1])$ such as all points of the interval $(-1, 0)$ are wandering. The other points of the closed interval $[-1, 1]$ are nonwandering, i.e.,

$$\Omega(\psi) = \{-1\} \cup ([0, 1] \setminus \bigcup_{n=0}^{+\infty} \psi^{-n}((y_0, 1]). \tag{6.21}$$

Let E be a nonempty closed subset of the interval $[0, 1]$. Define "Urysohn's hat" $h \in C^1([0, 1])$ such that $h(x) = 0$ for $x \in E$, and $0 < h(x) < 1$ for $x \in [0, 1] \setminus E$.

Example 2. Let $E = \{0\}$, $F_2(x, y) = (x, g_x(y))$, where $I = [0, 1] \times [-1, 1{,}01]$,

$$g_x(y) = \begin{cases} \psi(y), & \text{if } (x, y) \in [0, 1] \times [-1, 1); \\ h(x)(y-1)^2, & \text{if } (x, y) \in [0, 1] \times [1, 1{,}01]. \end{cases} \tag{6.22}$$

Then by (6.5), (6.21), and (6.22), the equality

$$\Omega(F_2) = \bigcup_{x\in[0,1]} \{x\} \times (\{1\} \cup \Omega(\psi))$$

holds; in addition, $1 \notin \Omega(\tilde{g}_0)$, $(0; 1) \in \overline{\bigcup_{x\in[0,1]} \{x\} \times \Omega(\psi)}$. This means that $y = 1$ is the weakly nonwandering point of the family of fiber maps for the skew product F_2.

We would like to point out that in [10, 12], one can find other examples of skew products of maps of an interval with properties similar to those of the above map F_2.

Example 3. Let E be the Cantor discontinuum on the interval $[0, 1]$. Define the skew product $F_3 \in T^1([0, 1] \times [-1, 1{,}01])$, where

$$F_3(x, y) = (x + h(x), g_x(y)),$$

and the fiber maps $g_x : [-1, 1{,}01] \to [-1, 1{,}01]$ are defined by (6.22) for all $x \in [0, 1]$. Then $Per(f) = Fix(f) = E$, where $Fix(\cdot)$ is the set of fixed points of a map; $(x; 1) \notin \zeta^{F_3|E\times I_2}$, but $(x; 1) \in \underset{n\to+\infty}{Ls}\, \eta^{ex'}_{l_n,1|U_{1,\varepsilon_n}(E)}$, and $(x; 1) \in \zeta^{F_3}$ for every $x \in E$. Hence, $y = 1$ is a weakly nonwandering point of the family of fiber maps for the skew product F_3 (compare with the examples introduced in [19]).

Finally, we will prove the second part [i.e., the equality (6.6)] of Theorem 1.

Lemma 4. *Let a map $F \in T^0(I)$ have the quotient map f with a closed set $Per(f)$. Let a periodic point x of f with (least) period $l(x)$ be a limit point for f-aperiodic points. Then*

$$(\zeta^F)(x) = \underset{n\to+\infty}{Ls}\, \eta^{ex'}_{l(x)n,1|U_{1,\varepsilon_{l(x)n}}}(x)$$

for an arbitrary sequence of positive numbers $\{\varepsilon_{l(x)n}\}_{n\geq 1}$ such that $\lim_{n\to+\infty} \varepsilon_{l(x)n} = 0$.

Proof. Let x be a periodic point of f with (least) period $l(x)$, and let x be a limit point for aperiodic points of f. Next, we will prove the following inclusion:

$$\underset{n\to+\infty}{Ls}\, \eta^{ex'}_{l(x)n,1|U_{1,\varepsilon_{l(x)n}}}(x) \subset (\zeta^F)(x). \tag{6.23}$$

Take an arbitrary point $y \in I_2$. The claim of Lemma 4 holds if $y \notin \Omega(\tilde{g}_x)$, but $(x; y) \in \zeta^{F|_{Per(f)\times I_2}}$ (for this case (i), see Lemma 3). Consider the case (ii), in which the point $y \in I_2$ satisfies the following condition: $y \notin (\zeta^{F|_{Per(f)\times I_2}})(x)$, but $(x; y) \in \underset{n\to+\infty}{Ls}\, \eta^{ex'}_{l(x)n,1|U_{1,\varepsilon_{l(x)n}}}(x)\backslash Per(f)$ for every sequence of positive numbers $\{\varepsilon_{l(x)n}\}_{n\geq 1}$ such that $\lim_{n\to+\infty} \varepsilon_{l(x)n} = 0$.

In this case, it is sufficient to prove that for every neighborhood $U_\varepsilon((x; y))$ of the point $(x; y)$ in I, there are a natural number $n = n(\varepsilon)$ and a point $(x_n; y_n)$ such that for some $r \geq 1$, $r = r((x_n; y_n))$, the following is valid:

$$(x_n; y_n),\ F^{l(x)nr}(x_n; y_n) \in U_\varepsilon((x; y)). \tag{6.24}$$

Let x be a nonrepelling f-periodic point. In this case, one can choose a neighborhood $U_1(x)$ of the point x in I_1 satisfying the inclusion

$$f^{l(x)}(U_1(x)) \subseteq U_1(x)\ \ [30]. \tag{6.25}$$

Let ε be an arbitrary positive number so small that $U_{1,\varepsilon/3}(x) \subset U_1(x)$ for the $\varepsilon/3$-neighborhood $U_{1,\varepsilon/3}(x)$ of the point x in I_1.

Use the continuity of the map $F^{l(x)}$ at the point $(x; y)$ and choose $\delta > 0$ (for $\varepsilon/3$) such that for every point $(x'; y') \in I$ satisfying $|x - x'|,\ |y - y'| < \delta$, the inequalities hold.

$$|x - f^{l(x)}(x')|,\ |g_{x,l(x)}(y) - g_{x',l(x)}(y')| < \varepsilon/3 \tag{6.26}$$

Use the neighborhood $U_\theta((x;y)) = U_{1,\theta}(x) \times U_{2,\theta}(y)$ of the point $(x;y)$ in I, where $\theta = \varepsilon/3$ if there is $\delta \geq \varepsilon/3$ with the above property, and $\theta \leq \delta$ if every $\delta > 0$ (for $\varepsilon/3$), by the continuity of the map $F^{l(x)}$ at the point $(x;y)$, satisfies the inequality $\delta < \varepsilon/3$.

Since $(x;y) \in \underset{n\to+\infty}{Ls}\ \eta^{ex'}_{l(x)n,1\,|U_{1,\varepsilon_{l(x)n}}(x)}$, it follows that in the case considered, the neighborhood $U_\theta((x;y))$ intersects the countable set of graphs of functions $\eta^{ex'}_{l(x)n,1\,|U_{1,\varepsilon_{l(x)n}}(x)}$ in points each of which does not belong to the fiber $\{x\} \times I_2$.

Take an arbitrary point

$$(x'_n; y'_n) \in U_\theta((x;y)) \bigcap \eta^{ex'}_{l(x)n,1\,|U_{1,\varepsilon_{l(x)n}}(x)}, \quad \text{where } x'_n \neq x.$$

By (6.3), we have $y'_n \in \Omega(g_{x''_n,\,l(x)n})$ for $x''_n = f^{l(x)n}(x'_n)$. These formulas with (6.25) imply $(x''_n; y'_n) \in U_\theta((x;y))$. Using the property 6 of Sect. 6.2, we find the point $y_n \in U_{2,\theta}(y)$ and the natural number r such that

$$g^r_{x''_n,\,l(x)n}(y_n) = y'_n. \tag{6.27}$$

Suppose that for $\varepsilon/3$, there exists a positive number $\delta \geq \varepsilon/3$ such that for every point $(x';y') \in I$ satisfying $|x - x'|,\ |y - y'| < \delta$, the inequalities (6.26) hold. In this case, we use the neighborhood $U_\theta((x;y))$ for $\theta = \varepsilon/3$.

Take an arbitrary point $x_n \in U_{1,\varepsilon/3}(x)$, $x_n \neq x$. Since $\delta \geq \varepsilon/3$, the sequence $\{g_{x_n,\,l(x)nj}(y_n)\}_{j\geq 0}$ of the second coordinates of the points $\{F^{l(x)nj}(x_n;y_n)\}_{j\geq 0}$ of the $F^{lx)n}$-trajectory of the point $(x_n;y_n)$ shadows up to $\varepsilon/3$ the trajectory $\{g^j_{x''_n,\,l(x)n}(y_n)\}_{j\geq 0}$ of the point y_n with respect to the map $g_{x''_n,\,l(x)n} : I_2 \to I_2$. In particular, we have for $j = r$,

$$|g^r_{x''_n,\,l(x)n}(y_n) - g_{x_n,\,l(x)nr}(y_n)| = |y'_n - g_{x_n,\,l(x)nr}(y_n)| < \varepsilon/3.$$

This last inequality, together with (6.25), immediately implies (6.24). Thus, we have in this case $(x;y) \in \Omega(F)$, and the inclusion (6.23) is valid.

Suppose that for $\varepsilon/3$, every $\delta > 0$ such that the inequalities (6.26) are satisfied (for every $(x';y')$ satisfying $|x - x'|,\ |y - y'| < \delta$) satisfies the inequality $\delta < \varepsilon/3$. Next, we will describe the choice of the neighborhood $U_\theta((x;y))$ for $\theta \leq \delta$ such that the special segments of negative subtrajectories of points from $U_\theta((x;y))$ possess the shadowing property.

In fact, since $(x;y) \in \underset{n\to+\infty}{Ls}\ \eta^{ex'}_{l(x)n,1\,|U_{1,\varepsilon_{l(x)n}}(x)}$ and formulas (6.25), (6.27) are satisfied, we have

$$(x;y) \in [\alpha'_1, \beta'_1] \times [\alpha'_2, \beta'_2], \quad \text{where } [\alpha'_1, \beta'_1] \times [\alpha'_2, \beta'_2] \subset \bigcap_{n=1}^{+\infty} F^{l(x)n}(\overline{U}_1(x) \times I_2)$$

for some $\alpha_1', \beta_1' \in \overline{U}_1(x)$, $\alpha_2', \beta_2' \in I_2$. Since $\delta < \varepsilon/3$, then $[\alpha_2', \beta_2']$ is a nondegenerate interval (an interval $[\alpha_1', \beta_1']$ can degenerate to the point x if x is an attracting fixed point of the map $f^{l(x)}$). Set θ equal to the least positive number from the set of three nonnegative numbers $\{\delta, y-\alpha_2', \beta_2'-y\}$.[3]

From (6.27), we have that the segment of length r of the negative semitrajectory of the point y_n' with respect to the map $g_{x_n'', l(x)n}$ is well defined, where $y_n \in g^{-r}_{x_n'', l(x)n}(y_n')$, $g^{-r}_{x_n'', l(x)n}(\cdot)$ is the complete preimage of a point. Set $\hat{x}_n = f^{l(x)nr}(x_n')$. $\hat{y}_n = y_n'$. Then $(\hat{x}_n; \hat{y}_n) \in U_\theta((x;y))$. Since $\theta \le \delta < \varepsilon/3$, it follows that the segment of length r of the negative semitrajectory of the point $\hat{y}_n$ with respect to the map $g_{x_n'', l(x)n} : I_2 \to I_2$ approximates up to δ the set of the points $\{\hat{y}_n^{-1}, \dots \hat{y}_n^{-r}\}$, where $g_{x_n', l(x)nj}(\hat{y}_n^{-j}) = \hat{y}_n$ for any $1 \le j \le r$. From this we conclude that

$$(x_n'; \hat{y}_n^{-r}),\ F^{l(x)nr}(x_n'; \hat{y}_n^{-r}) \in U_\varepsilon((x;y)).$$

Therefore, the inclusion (6.23) is satisfied by (6.24).

As above, let $x \in Per(f)$ have (least) period $l(x)$. Let any neighborhood of the point x contain aperiodic points of f. Take an arbitrary point

$$(x;y) \notin \underset{n\to+\infty}{Ls}\ \eta^{ex'}_{l(x)n,1\,|U_{1,\varepsilon_{l(x)n}}(x)}.$$

Using the definition of the multifunctions $\eta^{ex'}_{l(x)n,1\,|U_{1,\varepsilon_{l(x)}}(x)}$ and the considerations of the proof of Lemma 3, consider a neighborhood $U((x;y)) = U_{1,\varepsilon}(x) \times U_{2,\varepsilon}(y)$ $(0 < \varepsilon \le \varepsilon_{l(x)})$ of the point $(x;y)$ in I such that

$$U_{2,\varepsilon}(y) \bigcap g^j_{x', l(x)n}(U_{2,\varepsilon}(y)) = \emptyset \qquad (6.28)$$

for every $j \ge 1$, $n \ge 1$, and $x' \in U_{1,\varepsilon}(x)$; in addition, $U_{1,\varepsilon}(x) \bigcap f^k(U_{1,\varepsilon}(x)) \neq \emptyset$ iff $l(x)$ divides k (see property 2, Sect. 6.2). Setting $j=1$ in (6.28), we obtain that the equality $U_{2,\varepsilon}(y) \bigcap g_{x', l(x)n}(U_{2,\varepsilon}(y)) = \emptyset$ is valid for every $n \ge 1$ and $x' \in U_{1,\varepsilon}(x)$. The above, together with the choice of the neighborhood $U_{1,\varepsilon}(x)$, means that $(x;y) \notin \Omega(F)$. Hence, $(\zeta^F)(x) \subset \underset{n\to+\infty}{Ls}\ \eta^{ex'}_{l(x)n,1\,|U_{1,\varepsilon_{l(x)n}}(x)}$. This inclusion, together with (6.23), completes the proof of Lemma 4. □

And thus the proof of Theorem 1 is complete. □

We would like to point out that in contrast to to the classical theorem about families of ε-trajectories (see [1]), our proof of Lemma 4 does not require the hyperbolicity type for fiber maps.

Note also that one can substitute in Theorem 1 auxiliary multifunctions η'_{l_n} and the multifunctions introduced from them $\eta'_{l_n,1}$, $\eta^{ex'}_{l(x)n,1}$ with suitable multifunctions $\overline{\eta}'_{l_n}$ (for the Ω-function) and the multifunctions $\overline{\eta}'_{l_n,1}$, $\overline{\eta}^{ex'}_{l(x)n,1}$ derived from them.

---

[3] If $y-\alpha_2' = 0$ $(\beta_2' - y = 0)$. Then the neighborhood $U_2(y)$ is the right-hand (left-hand) neighborhood of the point y.

Completing this part of the work, we give an example of a C^1-smooth skew product of maps of an interval with closed set of periodic points such that every point of its phase space is chain recurrent, but the nonwandering set (coinciding with the periodic points set) is a polygonal line consisting of three components.

Definition 4 (See [2]). A point $z \in I$ is said to be a *chain-recurrent point for the map* $F \in T^0(I)$ if for every $\varepsilon > 0$, there exists a ε-chain leading from z to z with respect to the map $F \in T^0(I)$. Moreover, a finite set of points $\{u_k\}_{k=0}^n$ such that

$$u_0 = z_1,\ u_n = z_2, \text{ and } d(F(u_{k-1}), u_k) < \varepsilon \text{ for } k = 1, \dots, n,$$

where d is the metric in I conforming with the direct topology in I, is said to be an *ε-chain, connecting points z_1 and z_2* with respect to the map $F \in T^0(I)$.

Example 4 (See [6]). Consider the C^1-smooth skew product $F_4 : [0, 1]^2 \to [0, 1]^2$ defined by

$$F_4(x, y) = (x, y + 0,1(1 - x)(\cos 2\pi y - 1)).$$

Then the map F_4 has a closed set of periodic points satisfying

$$Per(F_4) = Fix(F_4) = L, \text{ where } L = ([0, 1] \times \{0; 1\}) \cup (\{1\} \times [0, 1]).$$

We obtain from this that $Per(F_4)$ is a closed set; in addition, $Per(g_x) = Fix(g_x) = \{0; 1\}$ for $x \in [0, 1)$, and $Per(g_1) = Fix(g_1) = [0, 1]$. It is not difficult to see that

$$CR(F_4) \setminus Per(F_4) = [0, 1]^2 \setminus L,$$

where $CR(\cdot)$ denotes the set of chain recurrent points of a map, and at the same time, $\Omega(F_4) = L$.

6.4 Nonwandering Set of C^1-Smooth Skew Products with a Closed Set of Periodic Points

In [18] is given an example of a continuous skew product of maps of the interval with a closed set of periodic points such that this map has nonwandering aperiodic points. Next, we will prove the impossibility of this situation for C^1-smooth skew products of interval maps.

Theorem 2. *If $F \in T^1(I)$, then the following claims are equivalent:*

(2.1) $\Omega(F) = Per(F)$.
(2.2) *The set of F-periodic points $Per(F)$ is closed.*

Recall that the C^1-norm of an arbitrary map $F \in T^1(I)$ is defined by the formula

$$||F||_1 = \max\{||F||_0, ||DF||_0\}, \tag{6.29}$$

where

$$||F||_0 = \max\{\sup_{x\in I_1}|f(x)|,\ \sup_{x,y\in I}|g_x(y)|\}$$

and

$$||DF||_0 = \max\{\sup_{x\in I_1}|f'(x)|,\ \sup_{x,y\in I}(|\frac{\partial g_x(y)}{\partial x}|+|\frac{\partial g_x(y)}{\partial y}|)\},$$

where $DF : I \to I$ is the differential of F.

We use the space $C^1(I_k)$ $(k = 1, 2)$ of all C^1-smooth maps of the interval I_k into itself with C^1-norm

$$||f||_{1,k} = \max\{\sup_{t\in I_k}|f(t)|, \sup_{t\in I_k}|f'(t)|\}$$

and the space $C^1(I, I_2)$ of all C^1-smooth maps of I into I_2 with C^1-norm

$$||g||_{1,(1,2)} = \max\{||g||_{0,(1,2)},\ ||Dg||_{0,(1,2)}\}, \tag{6.30}$$

where

$$||g||_{0,(1,2)} = \sup_{(x;y)\in I}|g(x,y)|$$

and

$$||Dg||_{0,(1,2)} = \sup_{(x,y)\in I}(|\frac{\partial g(x,y)}{\partial x}|+|\frac{\partial g(x,y)}{\partial y}|),$$

with $Dg : I \to I_2$ the differential of the map $g \in C^1(I, I_2)$. From (6.29) and (6.30), we get the inequality

$$\max\{||f||_{1,1},\ ||g_x||_{1,2}\} \le \max\{||f||_{1,1},\ ||g_x||_{1,(1,2}\} = ||F||_1 \tag{6.31}$$

for every $x \in I_1$. Associated with a skew product of maps of an interval $F \in T^1(I)$ there is a functional map $\rho_1 : I_1 \to C^1(I, I_2)$ (ρ_1 is called the C^1-representation) such that

$$\rho_1(x) = g_x \text{ for all } x \in I_1.$$

We say that the C^1-representation $\rho_1 : I_1 \to C^1(I, I_2)$ is *continuous* at a point $x' \in I_1$ if for every $\varepsilon > 0$ there exists a positive number $\delta = \delta(x', \varepsilon)$ such that the inequality

$$||g_x - g_{x'}||_{1,(1,2)} < \varepsilon$$

is valid for every $x \in I_1$ satisfying $|x - x'| < \delta$.

Note that $F \in T^1(I)$ iff $f \in C^1(I_1)$, and $\rho_1 : I_1 \to C^1(I, I_2)$ is continuous on I_1 (i.e., ρ_1 is continuous at every point of this interval in the above sense) [25].

Lemma 5. *If the set $Per(F)$ of periodic points of a skew product $F \in T^1(I)$ is closed, then $\Omega(F) = Per(F)$.*

Proof. Since $F \in T^1(I)$, then $\tau(F) = \{1, 2, \ldots 2^\nu\}$ for some $\nu < +\infty$ (where $\tau(F)$ is the set of the (least) periods of F-periodic points) [6].[4]

Let $\nu = 0$, i.e., $\tau(F) = \{1\}$. From [26], we know that there is an ε_1-neighborhood $B^1_{1,\varepsilon_1}(f)$ of the map f in $C^1(I_1)$ such that for every $\varphi \in B^1_{1,\varepsilon_1}(f)$, the inclusion $\tau(\varphi) \subseteq \{1; 2\}$ holds; for every $n \geq 2$ and $x \in Per(f)$, there is an $\varepsilon_n(x)$-neighborhood $B^1_{2,\varepsilon_n(x)}(g^n_x)$ of a map $g^n_x = g_{x,n}$ in $C^1(I_2)$ such that for every map $\theta \in B^1_{2,\varepsilon_n(x)}(g^n_x)$, the inclusion $\tau(\theta) \subseteq \{1; 2\}$ is valid.

For all $n \geq 2$ and $x \in Per(f)$, we use neighborhoods $B^1_{(1,2),\varepsilon_n(x)}(g^n_x)$ of C^1-smooth (with respect to the union of the variables x and y) fiber maps $\{g_{x,n}\}_{x \in I_1}$ of F^n. In particular, by the inequality (6.31), the inclusion

$$B^1_{(1,2),\varepsilon_n(x)}(g^n_x) \subset B^1_{2,\varepsilon_n(x)}(g^n_x)$$

holds for every $x \in Per(f)$ ($Per(f) = Per(f^n)$). Therefore, $\tau(g_{x',n}) \subseteq \{1; 2\}$ for every $x' \in I_1$ such that $g_{x',n} \in B^1_{(1,2),\varepsilon(x)}(g^n_x)$.

Let us use the C^1-representation $\rho_n : I_1 \to C^1(I, I_2)$, where $\rho_n(x) = g_{x,n}$ for every $x \in I_1$, $n \geq 2$. The continuity of ρ_n and the compactness of the set $Per(f)$ imply the compactness of the image $\rho_n(Per(f)) = \{g_{x,n}\}_{x \in Per(f)}$ in the family of C^1-smooth fiber maps $\{g_{x,n}\}_{x \in I_1}$ of F^n. The family of neighborhoods $\{B^1_{(1,2),\varepsilon_n(x)}(g^n_x)\}_{x \in Per(f)}$ is the open cover of the compact $\rho_n(Per(f))$. This cover is infinite iff $Per(f)$ is an infinite set. If the set $Per(f)$ is infinite, then we distinguish a finite subcover $\{B^1_{(1,2),\varepsilon_n(x_i)}(g^n_{x_i})\}_{i=1}^{r(n)}$ from the open cover $\{B^1_{(1,2),\varepsilon_n(x)}(g^n_x)\}_{x \in Per(f)}$ of the compact $\rho_n(Per(f))$.

Set $\varepsilon_n = \min\limits_{1 \leq i \leq r(n)} \{\varepsilon_n(x_i)\}$. By the uniform continuity of the C^1-representation $\rho_n, I_1 \to C^1(I, I_2)$, for $\varepsilon_n > 0$ there is a positive number $\delta_n < \varepsilon_n/2$ such that for every $x, x' \in I_1$ satisfying $|x - x'| < \delta_n$, the inequality $||g_{x,n} - g_{x',n}||_{1,(1,2)} < \varepsilon_n/2$ holds.

Let us fix an arbitrary number $0 < \delta'_n < \delta_n$ for every $n \geq 2$. Denote by $\overline{U}_{1,\delta'_n}(Fix(f))$ the closure of the δ'_n-neighborhood of the set $Fix(f)$ in I_1 and consider the multifunctions $\eta^{ex}_n{}_{|\overline{U}_{1,\delta'_n}(Fix(f))}$ $(n \geq 2)$. By the above, the graph of each of these functions coincides with the closed set of periodic points of the auxiliary skew product $\tilde{F}_n = (\tilde{f}, g_{x,n})$, where $\tilde{f} : I_1 \to I_1$ is the extension of the identity map from the closed set $\overline{U}_{1,\delta'_n}(Fix(f))$ on the interval I_1 such that the set of all periodic points of $\tilde{f}$ [here $Per(\tilde{f}) = Fix(\tilde{f})$] coincides with $\overline{U}_{1,\delta'_n}(Fix(f))$. Using the closedness of the

---

[4] An analogous result can be found in [7], in which is stated the existence of a C^∞-smooth skew product of maps of an interval of type $\prec 2^\infty$ with one-dimensional attracting set. In [7], the skew product is realized as the shift map along the trajectories (defined for every t) of the corresponding nonautonomous system of differential equations with C^∞-smooth right-hand sides. This means that considerations are relative to $\mathbf{R}^3$, and oscillations of the trajectory in a neighborhood of its limit set are "distributed" along the unbounded axis t. In the consideration of a skew product in a rectangle of the plane xOy, there is no opportunity to "distribute" oscillations of a trajectory possessing a one-dimensional attracting set. As a result, this leads to oscillations of the partial derivative $\frac{\partial}{\partial x} g_x(y)$ and its unboundedness in a neighborhood of the attracting set, although the map $g_x(y)$ can be a C^∞-map with respect to y (but not with respect to the union of the variables x and y) [13].

set $Per(\tilde{F}_n)$, the compactness of I_2, and the property that I_1 is a Hausdorff space, we obtain that every multifunction $\eta_n^{ex}{}_{|\overline{U}_{1,\delta'_n}(Fix(f))}$ $(n \geq 2)$ is upper semicontinuous.

Take an arbitrary point $x \in \Omega(f) = Fix(f)$ and fix it. By the upper semicontinuity of $\eta_n^{ex}{}_{|\overline{U}_{1,\delta'_n}(Fix(f))}$, for a number $\varepsilon_n > 0$, $n \geq 2$, there is $0 < \delta''_n < \delta'_n$ such that for all $x' \in I_1$ satisfying $|x' - x| < \delta''_n$, the inclusion $\eta_n^{ex}(x') \subset U_{2,\varepsilon_n}(Per(\tilde{g}_x))$ holds, where $U_{2,\varepsilon_n}(Per(\tilde{g}_x))$ is an ε_n-neighborhood of the set $Per(\tilde{g}_x)$ in I_2.

Choose a sequence $\{\varepsilon_n\}_{n\geq 2}$ such that $\lim_{n\to\infty} \varepsilon_n = 0$. Then using the continuity of f and formula (6.3), and applying Theorem 1 [see the equality (6.6)], we obtain $\zeta^F(x) \subset Per(\tilde{g}_x)$, from which we get the desired equality (2.1).

Let $\nu \geq 1$. Then $\Omega(F^{2^\nu}) = Per(F)$ (see step 1). By Definition 1, the inclusion $\Omega(F^{2^\nu}) \subset \Omega(F)$ is valid. We will prove that in this case, the inverse inclusion

$$\Omega(F) \subset \Omega(F^{2^\nu}) \tag{6.32}$$

holds. In fact, take an arbitrary point $(x^0; y^0) \in \Omega(F)$. By property 2 of Sect. 6.2, we immediately have $x^0 \in Per(f)$.

Denote by $n(x^0)$ the (least) period of the point x^0. Since $Per(f)$ is a closed set, there is a neighborhood $U_1(x^0) \subset I_1$ of the point x^0 satisfying $U_1(x^0) \bigcap f^n(U_1(x^0)) \neq \emptyset$ iff n is a multiple of the (least) period $n(x^0) = 2^{i_0}$ of the point x^0 [27]. Therefore, using (6.1), we obtain that $(x^0; y^0) \in \Omega(F)$ iff $(x^0; y^0) \in \Omega(F^{n(x^0)})$. Hence, $(x^0; y^0) \in \Omega(F^{2^\nu})$ for $n(x^0) = 2^\nu$.

Let $n(x^0) = 2^{\nu-1}$. If the point $(x^0; y^0)$ is a limit point of the set $\Omega(F)$ and x^0 is a limit point of a sequence of points from $Per(f)$ such that the least period of each of them equals 2^ν, then using the closedness of the set $\Omega(F)$, we obtain from the above that $(x^0; y^0) \in \Omega(F^{2^\nu})$. Suppose that $(x^0; y^0)$ does not possess the above property. In this case, either x^0 is an isolated point of the set $Per(f)$ or x^0 is a limit point for the sequence of points from $Per(f)$ such that the least period of each of them equals $2^{\nu-1}$ (but not 2^ν). Therefore, there exists a neighborhood $U_1(x^0)$ of the point x^0 in I_1 that does not contain points of the set $Per(f)$ with least period 2^ν. Repeating the arguments of step 1 for $(f_{|\overline{U_1}(x^0)})^{2^\nu}$, we verify that $(x^0; y^0) \in \Omega(F^{2^\nu})$.

Repeating the above arguments successively for $n(x^0) = 2^{\nu-p}$, where $2 \leq p \leq \nu - 1$, we verify that $(x^0; y^0) \in \Omega(F^{2^\nu})$. Hence, the inclusion (6.32) is valid. This inclusion, together with the inverse inclusion proves (2.1), and Lemma 5 is proved. □

Since the nonwandering set $\Omega(F)$ is closed, then, (2.1) (without assumptions under the smoothness of the dynamical system) implies (2.2), and Theorem 2 is proved. □

Acknowledgements The Author is partially supported by the Federal Target Program "Scientific and Scientific-Pedagogical Personnel of Innovative Russia" (2009–2013) of the Education Ministry of Russia, grant No 14.B37.21.0361.

References

1. Anosov, D.V.: On one class of invariant sets of smooth dynamical systems [in Russian]. In: Metropol'skii, Yu.A. (ed.) Proceedings of V International Conference on Nonlinear Oscillations, vol. 2, pp. 39–45. Institute of Mathematics of Academy of Sciences of the Ukrainian SSR, Kiev (1970)
2. Arnold, V.I., Ilyashenko, Yu.S., Anosov D.V.: Dynamical systems – 1, book 1: ordinary differential equations [in Russian]. In: Gamkrelidze, R.V. (ed.) Itogi nauki i tekhniki. Sovremennye problemy matematiki. Fundamental'nyye napravleniya, vol. 1. VINITI, Moscow (1985)
3. Arteaga, C.: Smooth triangular maps of the square with closed set of periodic points. J. Math. Anal. Appl. **196**, 987–997 (1995)
4. Beck C. : Chaotic cascade model for turbulent velocity distributions. Phys. Rev. **E 49**, 3641–3652 (1994)
5. Bel'mesova S.S., Efremova L.S.: On invariant sets of some quadratic maps of the plane. Vestnik of Nizhni Novgorod State University named after N.I. Lobachevskii [in Russian] **2**, 152–158 (2012)
6. Blinova, E.V., Efremova, L.S.: On Ω-blow ups in the simplest C^1-smooth skew products of maps of an interval. J. Math. Sci. (N.Y.) **157**, 456–465 (2009)
7. Bruno, D., Lopez, V.J.: Asymptotical periodicity for analytic triangular maps of type less than 2^∞. J. Math. Anal. Appl. **361**, 1–9 (2010)
8. Coven, E.M., Nitecki, Z.: Nonwandering sets of powers of maps of the interval. Ergod. Theory Dyn. Syst. **1**, 9–31 (1981)
9. Davies M.E., Campbell K.M.: Linear recursive filters and nonlinear dynamics. Nonlinearity **9**, 487–499 (1996)
10. Efremova, L.S.: On the nonwandering set and the center of triangular mappings with a closed set of periodic points in the base [in Russian]. In: Sharkovskii, A.N. (ed.) Dynamical Systems and Nonlinear Phenomena, pp. 15–25. Institute of Mathematics of Academy of Sciences of the Ukrainian SSR, Kiev (1990)
11. Efremova, L.S.: On the Concept of Ω-function for a skew product of interval maps [in Russian]. In: Aseev, S.M., Vakhrameev, S.A. (eds.) Itogi nauki i tekhniki. Covremennaya matematika i ee prilozheniya, vol. 67, pp. 129–160. VINITI, Moscow (1999) [English translation, J. Math. Sci. (New York). **105**, 1779–1798 (2001)]
12. Efremova, L.S.: On the nonwandering set and the center of some skew products of interval maps. Russ. Math. (Iz.VUZ) **50**, 17–25 (2006)
13. Efremova, L.S.: Differential properties and attracting sets of a simplest skew product of interval maps. Sbornik: Math. **201**, 873–907 (2010)
14. Efremova, L.S.: Space of C^1-Smooth skew products of maps of an interval. Theory Math. Phys. **164**, 1208–1214 (2010)
15. Efremova, L.S.: The smooth skew product in the plane possessing ramified continuum as the global attractor. In: Proceedings of International Workshop on Nonlinear Maps and Their Applications (NOMA'11), pp. 31–33 (2011)
16. Fedorenko, V.V., Sharkovskii, A.N.: Continuous maps of an interval with a closed set of periodic points [in Russian]. In: Sharkovskii, A.N. (ed.) Investigations of Differential and Differential–Difference Equations, pp. 137–145. Institute of Mathematics of Academy of Sciences of the Ukrainian SSR, Kiev (1980)
17. Glendinning, P.: Global attractors of pinched skew products. Dyn. Syst. **17**, 287–294 (2002)
18. Guirao, J.L.G., Pelao, F.L.: On skew poduct maps with the base having a closed set of periodic points. Int. J. Comput. Math. **83**, 441–445 (2008)
19. Guirao, J.L.G., Rubio, R.G.: Nonwandering set of skew product maps with base having closed set of periodic points. J. Math. Anal. Appl. **362**, 350–354 (2010)
20. Jäger, T.H.: Quasiperiodically forced interval maps with negative Schwarzian derivative. Nonlinearity **16**, 1239–1255 (2003)
21. *Jäger*, T.H.: On the structure of strange nonchaotic attractors in pinched skew products. Ergod. Theory Dyn. Syst. **27**, 493–510 (2007)

22. Katok, A., Hasselblatt, B.: Introduction to the Modern Theory of Dynamical Systems. Cambridge University Press, Cambridge (1995)
23. Kolyada, S.F.: On dynamics of triangular maps of the square. Ergod. Theory Dyn. Syst. **12**, 749–768 (1992)
24. Kupka, J.: The triangular maps with closed sets of periodic points. J. Math. Anal. Appl. **319**, 302–314 (2006)
25. Kuratowski,K.: Topology. Academic, New York-London (1966)
26. Misiurewicz, M.: Structure of mappings of an interval with zero topological entropy. Inst. Hautes Études Sci. Publ. Math. **53**, 5–16 (1981)
27. Nitecki, Z.: Maps of the interval with closed periodic set. Proc. Am. Math. Soc. **85**, 451–456 (1982)
28. Nitecky, Z.: Topological dynamics on the interval. Prog. Math. **21**, 1–73 (1982)
29. Sharkovskii, A.N.: Cycles and the structure of a continuous map [in Russian]. Ukrain. Mat. Zh. **17**, 104–111 (1965)
30. Sharkovskii, A.N.: On one classification of fixed points [in Russian]. Ukr. Math. J. **17**, 80–95 (1965)
31. Stark, J.: Regularity of invariant graphs for forced systems. Ergod. Theory Dyn. Syst. **19**, 155–199 (1999)
32. Stark, J.: Transitive sets for quasiperiodically forced monotone maps. Dyn. Syst. **18**, 351–364 (2003)

Chapter 7
Green Route Planner

João Ferreira

Abstract This work is about the problem of multimodal shortest paths in transportation network systems where users have several modes and forms to travel from an origin to a destination. To extend the traditional shortest-path problem, an innovative framework is presented that integrates a multidecision criterion related to political environmental policies. The aim is to deal with an efficient design for the multimodal shortest-path computation taking into account not only the expected travel time, but also additional constraints such as real-time traffic information and CO_2 policies (e.g., a penalty is introduced based on CO_2 emissions).

7.1 Introduction

In the European Union, over 60% of the population lives in urban areas (information from Eurostat). Air and noise pollution is getting worse every year. Urban traffic is responsible for 40% of CO_2 emissions and 70% of emissions of other pollutants arising from road transportation [1]. Increasing traffic in town and city centers is responsible for chronic congestion, with the many adverse consequences that this entails in terms of delays and pollution. Every year, nearly 100 billion euros, or 1% of the EU's GDP (Gross Domestic Product) [1], is spent by the European economy to deal with this phenomenon. Several solutions have been proposed to these problems, such as a diversity of intelligent transportation systems and solutions.

Our transportation network system includes several travel modalities. Users have distinct alternatives for traveling from one place to another such as private car, car pooling, car or bike sharing, and public transportation (e.g., bus, train, underground, tram). The multimodal traveling alternatives proposed tousers seek to take

J. Ferreira (✉)
ISEL, Lisbon, Portugal
e-mail: jferreira@deetc.isel.ipl.pt

C. Grácio et al. (eds.), *Nonlinear Maps and their Applications*, Springer Proceedings in Mathematics & Statistics 57, DOI 10.1007/978-1-4614-9161-3_7,

advantage of the several networks by linking two or more different transportation modes to move passengers from a starting point to a destination point. A trip chain for user travel may include several modes. A work commute might involve driving a car to a railway station or bus stop, with the possibility of additional transfers within or among modes to complete the trip. A significant travel constraint in the multimodal context concerns the number of modal transfers that the user is able to execute in one origin–destination trip. Our proposal is based on this scenario and takes into account real-time traffic information to provide the best options, taking into account predefined criteria (such as fastest trip, cheapest trip, and also guidance). Finding such a best path with real-time traffic information with multidecision criteria is the subject of the present work. This tool (application) can be used by transportation policy-makers to come up with a workable penalization regime to reduce single-rider car use in big cities.

7.2 Related Work

Due to recent advances in computing, geographic information systems (GIS), mobile devices, communication and storage technologies, automatic transit trip planners have been implemented in recent years. Transit trip planners accept the origin, destination, and expected departure/arrival time as user input and find suitable routes using available transit services. Transit trip planners are generally web-based. Most existing trip planners are based on static schedule data. However, transit vehicles are often delayed by traffic congestion and accidents. In particular, transit buses are often late during peak-traffic hours in metropolitan areas. Trips based on static schedule data may make planned transfers infeasible if some transit vehicles are late. In the multimodal routing context, many works have studied the multimodal shortest viable path problem [2,3]. Three main problems are considered: (1) set of transit modes used to accomplish the desired path; (2) the number of modal transfers performed along the desired path; and (3) computational time to find the best solution in real time. Our approach uses Dijkstra's algorithm with several heuristics to speed up the calculation process. For example, a bike-sharing system is considered only if the total distance is less than 2 km (this is a user-configurable parameter). In addition, the system tries to avoid or minimize transportation transfers (see Sect. 7.6).

7.3 Mobi-System

The idea is to pass possible itineraries using public transportation to a graph, where the arc length is defined by the time that it takes to go from one node to the other. The same procedure is applied for car-sharing, carpooling, and bike-sharing systems. If we have all this information in a graph, we can use the best-path algorithm described in the next section. The big issue is matrix size, which can increase enormously when there is a great diversity of options, and that could generate computer memory problems in handling the matrix.

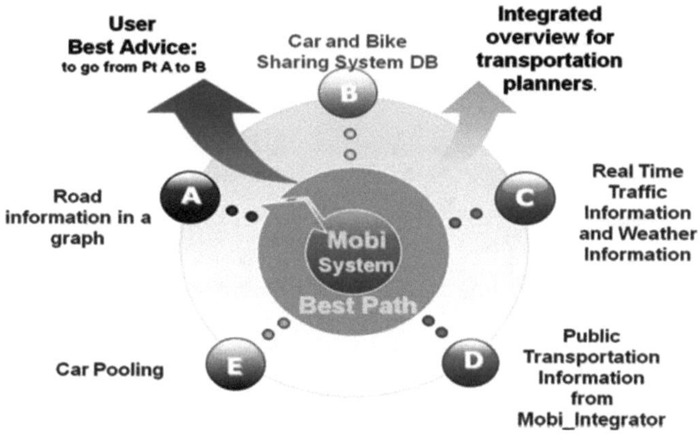

Fig. 7.1 Mobi-system transportation-related system

The arc weight can be constructed from the diversity of options, such as time and price. Also, this integrated approach to a diversity of systems with geographic information could be important for transportation planners or for transportation decision-makers. The Mobi-System is GIS-Technology integrated with spatial network modeling and promises to bring solutions to transit planning problems in mobile devices. A technological system description can be found at [4] and (Fig. 7.1).

7.4 Network Model

To represent the multimodal network, a structure composed of different unimodal interconnected layers is used (see Fig. 7.2). A multimodal path can then be viewed as a set of interconnected unimodal paths of public transportation, car-sharing, or bike-sharing. Given that the multimodal graph has been constructed and travel times and mode attributes have been assigned to each edge, we first propose a label-setting algorithm providing a multimodal route for a one-way trip between a starting point x and termination point y. This multimodal transportation network has transfer points previously defined by public transportation stops, parking places, or car- or bike-sharing stations that are within a 100-meter (configurable value) radius of a graph point. We define a label-setting algorithm giving an optimal one-way multimodal shortest path. Then, based on this algorithm, we build a strategy to solve the two-way problem, and we show how this strategy could be extended to solve the three-trip shortest multimodal chain problem.

Transport modes can be classified in two categories depending on whether they are subject to parking constraints. A unimodal path using an individual vehicle is said to be constrained, since it must begin at a node where a vehicle is available and end at a node where it is possible to park. The set of constrained modes is here reduced to a single element. The set of unconstrained modes contains all

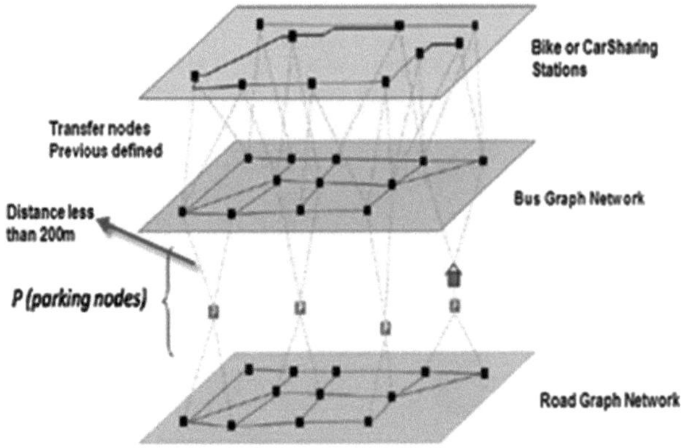

Fig. 7.2 Structure of the multimodal graph

other modes. All transfer points have computed transfer time; for example, parking a car is considered to take three minutes plus a walking time that is a function of the distance. All this information goes into an XML graph file.

7.5 Best Path

The main idea is to adapt the arc weight to a combination of items that might reflect an environmental policy. The arc weight is usually measure by the time that takes to go from start node (A) to the end node (B). Our approach is based on a new measure for this weight. So we propose as arc weight the travel time multiplied by a factor (we called price) that is related with CO_2 emissions. This factor is illustrated in Fig 7.7 (see price column) can be used to measure the impact for CO_2 emissions to go from node A to node B by the usage public transportation, private car, car-sharing, bike-sharing, or car-pooling [5] (the trip cost is divided by the number of persons in a car). This weight could include also a parameter function of city traffic conditions (overloaded paths should be more heavily penalized).

Real-time information is available from road concessionaires, but access to outside information is usually denied. To solve this problem in a project at ISEL [6], we created a web crawler to obtain traffic information from specialized sites, and from predefined heuristics an XML file with traffic information is created. The XML file is an approach for future data integration from different source providers. In Fig. 7.3 is illustrated the process of this XML file creation where we put together road information, GPS coordinates, distance and maximum speed. A similar process is performed for public transportation, where the stops represent the nodes.

We use *WebNews Crawler* (version 1.0), written by Vladimir Poroshin, configured to take traffic information from the TVI web site `http://www.tvi.iol.pt/`

`transito.php`. Information about nodes were checked against a heuristic table, where a conversion factor CVF will reduce the node speed traffic (if the rode is blocked, with no traffic flow, the CVF is zero). Figure 7.3 illustrates this process. The results are stored in an adjacency matrix, where the number a_{ij} represents the cost of going from i to j. To determine the best path between two points on a map, it creates a graph representing the map, where the arcs represent roads, and nodes represent intersections or traffic areas. It applies an algorithm to the graph to find the path with least weight and fastest time between the two desired points. The lead weight is determined based on the distance to us, at full speed, and a weighting factor was introduced based on real-time traffic information loaded into the system. The weight of an arc is basically the average required time in seconds. For this, the formula $\frac{L \times 3600}{V}$ is used, where L represents the size of the graph in kilometers, and V is the top speed. For all arcs, there is a speed limit, which serves as the basis for the weight when there is no traffic information. When there is traffic information for a particular arc, its weight is affected, because the degree of traffic congestion influences the speed. For example, if an arc with heavy traffic has a speed limit of 90 km/h given the existing conditions, the reduction factor with respect to the maximum speed will have a value set between 0 and 1, based on current traffic information, causing an increase in the time required for this arc to be traversed as soon as the weight is increased.

If the transit is cut off completely, the factor of speed reduction will have the value 0, and the maximum attainable speed, which is now 0 km/h causes this arc to have infinite weight.

Arc	Coordenates (lat,long)	Name
1	38.927410, -9.011100	Alhandra
2	38.894050, -9.048801	Alverca
3	38.848341, -9.089346	Santa Iria
4	38.790796, -9.114538	Sacavem
5	38.787406, -9.105337	Portela/Moscavide
6	38.717140, -8.948650	Montijo
7	38.740385, -8.931608	Alcochete
8	38.642651, -9.049044	Lavradio
9	38.607083, -9.019177	Nó IC21-IC32
10	38.583166, -9.017658	Coina
11	38.589418, -8.864434	Lau
...	...	...
50	38.689940, -9.177090	Pte 25 Abril
51	38.700951, -9.424789	Cascais
52	38.717284, -9.385188	Golf do Estoril

Arch	Road	Lenght (m)	Node Max. Speed (Km/h)
(1,2)	A1	5000	120
(2,3)	A1	6200	120
(3,4)	A1	7000	120
(3,5)	IC2	8500	80
(5,49)	IC2-A12	6600	120
(49,6)	A12	9900	120
(6,7)	IC3	4500	120
...	...	...	
(37,39)	A8-A9	11500	120
(39,40)	A9-A10	5800	120
(39,2)	A9-A1	3300	120


```
<graph>
    <node id="1" name="Alhandra" lat="38.927410" long="-9.011100"/>
    <node id="2" name="Alverca" lat="38.894050" long="-9.048801"/>
    <node id="3" name="Santa Iria" lat="38.848341" long="-9.089346"/>
    <node id="4" name="Sacavem" lat="38.790796" long="-9.114538"/>
    <node id="5" name="Portela" lat="38.787406" long="-9.105337"/>
    ...
    <node id="51" name="Cascais" lat="38.700951" long="-9.424789"/>
    <node id="52" name="Golf do Estoril" lat="38.717284" long="-9.385188"/>
    <node id="53" name="Pontinha" lat="38.768703" long="-9.209198"/>
    <arc id="1" name="A1" node1="1" node2="2" length="5000" maxSpeed="120"/>
    <arc id="2" name="A1" node1="2" node2="3" length="6200" maxSpeed="120"/>
    <arc id="3" name="A1" node1="3" node2="4" length="7000" maxSpeed="120"/>
    <arc id="4" name="IC2" node1="3" node2="5" length="8500" maxSpeed="80"/>
    <arc id="5" name="A12" node1="5" node2="49" length="6600" maxSpeed="120"/>
    ...
    <arc id="68" name="A10" node1="39" node2="40" length="5800" maxSpeed="120"/>
    <arc id="69" name="A9" node1="39" node2="2" length="3300" maxSpeed="120"/>
    <arc id="70" name="Eixo Norte-Sul" node1="18" node2="17" lenght="5000" maxSpeed="90"/>
</graph>
```


Fig. 7.3 Semiautomatic creation of an XML file with road information to be used for graph creation

If traffic is proceeding smoothly and without any problems, the factor of speed reduction will have a value, and the maximum attainable speed becomes equal to the speed limit, which results in this arc having the lowest weight possible, translating into a reduction in the mean time required for the arc to be traversed. This weight can also integrate publically available transportation information. Users can choose the impact parameters based on their strategies, in order to save money, to satisfy conveniences, or even to save time. See the work presented in [6] for a complete description of this process.

Example 1. Path: A1 (Alhandra–Alverca). The speed limit on this section is 120 km/h (motorway), but traffic information shows an average speed of only 84 km/h. The distance to be traveled is 5 km, so the arc has weight

$$Aw = \frac{\text{arc length}}{\text{average speed}} \times 3600\,\text{s} = \frac{5000}{84000} \times 3600 = 0.0595238 \times 3600 \approx 214\,\text{s} \quad (7.1)$$

Let us now suppose that using public transportation on this route takes approximately twice as long, say 400 s. The transportation price is 1 euro, and the cost of transportation by automobile is 2 euros. If drivers weight the factors for time and price equally, then the new arc weights are $214 \times 2 = 428$ for travel by automobile, while the weight for public transportation remains at 400. In this case, these values are approximately the same, but if we add the cost of parking a car in the city, the arc weight for a private car will increase, and then the system will suggest public transportation option, giving the available table time information about nearby public transportation and also guidance for the corresponding local of public transportation stop. To determine the least expensive route or the quickest route was implemented in Dijkstra's algorithm, where its running time is $O\left(N^2\right)$, N being the number of nodes in the graph. The graph to be used by Dijkstra's algorithm is represented as an XML file. The best way of taking user preferences into account (e.g., whether to save time or money) will be determined using the Dijkstra algorithm. In our web application, there is a class called Dijkstra, which is where the algorithm is implemented. In our implementation, we used the following data structures: (1) an array of nodes (nodes) containing all nodes in the graph; (2) a list of integers *unsettledNodes*, which holds the ids of the nodes for which the best path to the source node has been found thus far; (3) a list of integers *settledNodes* that holds the best path to the source node found thus far; (4) a dictionary *shortestDistances*, which is the key to an integer representing the ids of the nodes; it has as its value another integer that represents the best way to estimate the source node; (5) a dictionary *predecessors*, which is the key to an integer representing the ids of nodes; it is another integer value, representing the ids of nodes in the previous best path to the source node. For a complete description of our method, see [6]. Information exchange with external systems is based on XML files.

7.6 Case Study

A different approach can be integrated into an arc-weight calculation of time, price, and weight related to CO_2 emissions. Figure 7.5 shows a configurable application screen where the weights (from 0 to 500) can be easily configured by

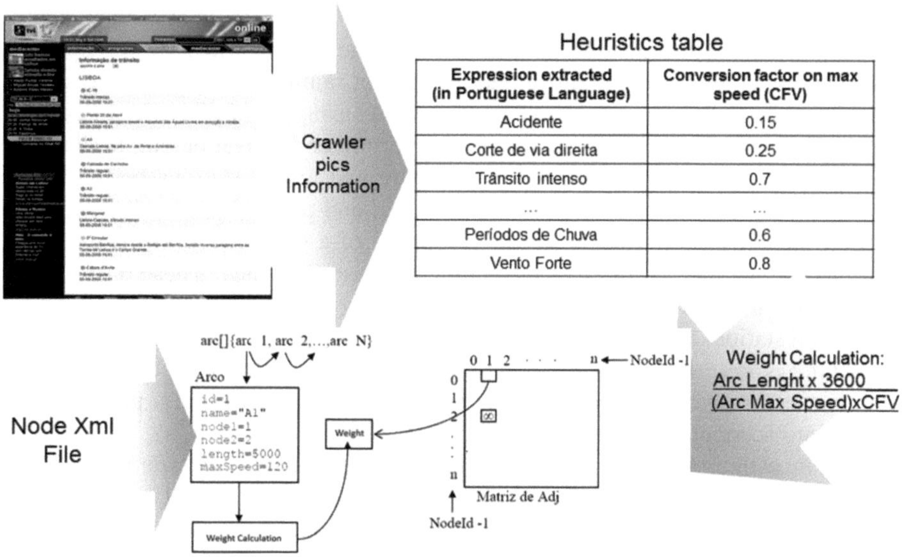

Fig. 7.4 Graph node weight calculation based on traffic information from web sites

the user. The best path is calculated based on predefined and configurable heuristics: (1) Begin with an estimate of distance between the start and end points. If this

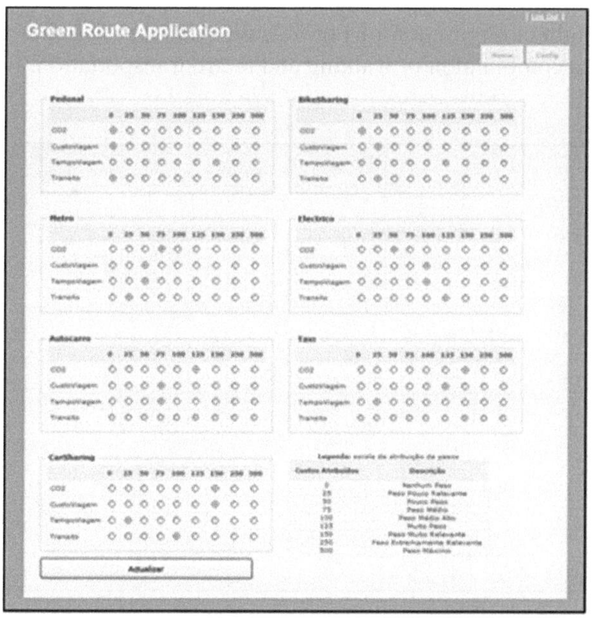

Fig. 7.5 Administration application menu: weight configuration

distance is less than 2 km (configurable parameter) and the altitude accumulated is less than 250 m (configurable parameter), then the system will look for a nearby bike-sharing station and suggest a bike if it is not raining. If this fails, the system calculates the paths we considered in two phases using the algorithm of Sect. 7.5. First it is checked whether the stations of departure and arrival belong to the same line (no transfer is necessary). If the departure and arrival stations belong to the same line *IsSameLine*, the field is marked true, and otherwise, it will be marked false. If *IsSameLine* is true, then the first phase of a prospective trip is executed, and thereby the data required to design the route are obtained. These data include the coordinates of the stations, the price of the trip, points of interest that are near the arrival station, and other information such as schedules and other features. If *IsSameLine* is false, then a second phase is run, in which the departure and arrival stations are not on the same line. It is therefore necessary to find a path or paths taking into account the available interfaces on the lines.

This investigation takes advantage of the method *getAllInterfacesBetween*, which was implemented in a recursive manner to allow the taking of all possible paths between two stations. To obtain the possible paths, an analysis was carried out taking into account the interfaces of the line of the station that is being verified. At each public transportation transfer point, it is checked whether the line matches the line of the arrival station? If so we store path, and if so, otherwise, it uses recursion to find a route.

At same time, we calculate the time for car transportation, and car sharing is proposed if the pick-up or drop-off station is within a distance of 500 m of the start and end points. To clarify this, we illustrate the following example, in which the user wants to go from Chelas to Bairro Alto in Lisbon; see Fig. 7.6. For this problem and the applied weight configuration of Fig. 7.7, the proposed solution will propose as the best option a combination of walking and metro transportation. If we change the

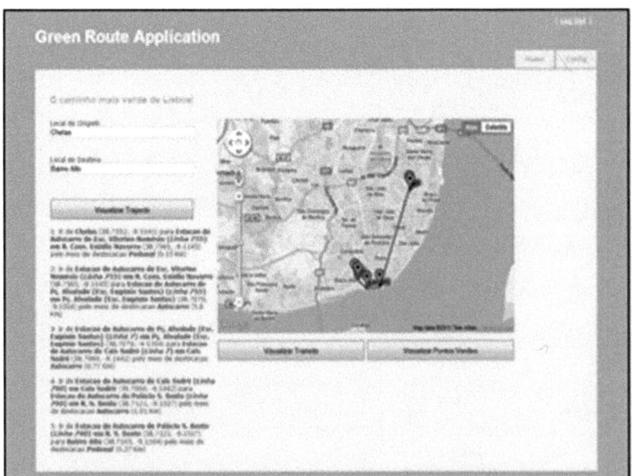

Fig. 7.6 Green route application

	CO_2 Emissions	Travel Time	Price	Weight
Pedestrian	0	150	0	150
Bike Sharing	0	125	1	200
Metro	50	75	2	225
Tram	50	100	2	300
Bus	75	100	2	325
Car Sharing	100	50	5	425
Taxi	100	50	15	400
Car	150	50	10	475

Fig. 7.7 Weights used on current example

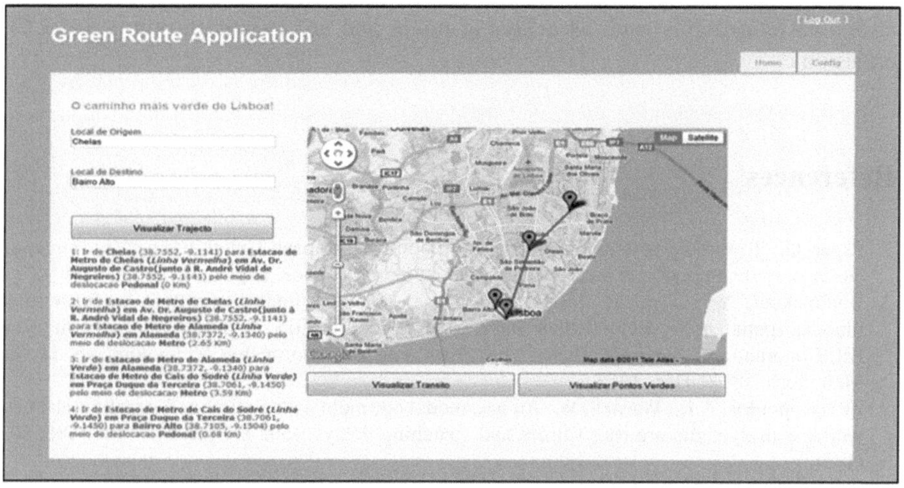

Fig. 7.8 Green route application results for all transportation with equal weights

weights (all equal), the best path will be based on the fastest option, based on bus transportation; see Fig. 7.8.

In Fig. 7.6, we have a first path of 200 m (walking), with $Aw_{\text{walk}} = 120$ s the walking time (more or less 2 min). Then we have metro transportation with $Aw_{\text{metro}} = 900 + 600 = 1500$ s. These values are taken from a standard timetable. The last path is again to walk 680 m, with $Aw_{\text{walk}} = 408$ s. Taking into account the weight for walking, 0.15 (see Fig. 7.7, table values divided by 1000), and the weight for metro, 0.225, we have $Aw_{\text{final}} = 508 \times 0.15 + 1500 \times 0.225 = 414$. For another possibility, see Fig. 7.8, with a total walking distance of 420 m, which gives $Aw_{\text{walk}} = 252$ s and with buses $Aw_{\text{bus}} = 1250$ s. Taking into account the bus weight of 0.325 (Fig. 7.7), we have $Aw_{\text{final}} = 252 \times 0.15 + 1250 \times 0.325 = 444$. In case of Fig. 7.7, the bus weight is changed to 0.225. Then $Aw_{\text{final}} = 252 \times 0.15 + 1250 \times 0.225 = 319$, so the second option is now the best.

7.7 Conclusion

Our transportation network system is based on several options (private car, car sharing, public transportation, walking, and bike-sharing). This system is viewed as a multimodal graph, and each mode network is modeled by a unimodal subgraph. Mode network structuring was used to speed up the computation of the multimodal shortest viable path and also to perform a new design for the multimodal path operator. The heuristics used to speed up the calculation process should be checked against several options. This study has examined a sustainable mobility approach toward a green environment policy. The present work is part of the Mobi-System, in which we propose a way of obtaining real-time traffic information and the introduction of CO_2 policies related to transportation emissions. The proposed algorithm takes into account not only the expected travel time, but also additional constraints such as delays at mode and arc switching points and CO_2 emissions.

References

1. Paper, G.: Towards a new culture for urban mobility. European Union. http://ec.europa.eu/transport/urban/urban_mobility/green_paper/green_paper_en.htm (2007)
2. Boulmakoul, A., Laurini, R., Mouncif, H., Taqafi, G.: Path-finding operators for fuzzy multimodal spatial networks and their integration in mobile GIS. In: Proceedings of the 2nd IEEE International Symposium on Signal Processing and Information Technology, pp. 51–56, Marrakech, 18–21 Dec. 2012
3. Ziliaskopoulos, A.K., Wardell, W.: An intermodal optimum path algorithm for multimodal networks with dynamic arc travel times and switching delays. Eur. J. Oper. Res. **125**, 486–502 (1998)
4. Ferreira, J.C., Afonso, J.L.: Mobi-system: a personal travel assistance for electrical vehicles in smart cities. In: 20th IEEE International Symposium on Industrial Electronics (ISIES 2011), Gdansk University of Technology, Poland, 27–30 June 2011
5. Ferreira, J.C., Trigo, P., Filipe, P.: Collaborative car pooling system. In: Proceedings of International Conference on Sustainable Urban Transport and Environment, Paris, 24–26 June 2009
6. Ferreira J.: MultiModal best path. In: Proceedings of the First Conference in Electronics, Telecommunications, and Computer Engineering (ISBN: 978-989-97531-0-5), at Instituto Superior de Engenharia de Lisboa-ISEL, Lisbon-Portugal. 24–25 November 2011

Chapter 8
Families of Hyperbolic Lorenz Knots

Paulo Gomes, Nuno Franco, and Luís Silva

Abstract Lorenz knots and links have been an area of research for over thirty years. Their study combines several fields of mathematics, including topology, geometry, and dynamical systems. The introduction of the Lorenz template permitted a dimension reduction, the definition of the one-dimensional Lorenz map, and the use of symbolic dynamics to code its orbits. Meanwhile, Thurston's geometrization theorem implies that all Lorenz knots can be classified as torus knots, satellites, or hyperbolic knots. Based on a list of hyperbolic knots presented by Birman and Kofman, we define three families of Lorenz knots, compute their hyperbolic volume using the topology and geometry software *SnapPy* and use the results to present conjectures concerning their hyperbolicity.

8.1 Introduction

Almost 50 years ago, Lorenz [8] introduced a simple model for convection in hydrodynamic systems, namely the atmosphere. This is a system of three ordinary differential equations that determine a flow in $\mathbf{R}^3$, the *Lorenz equations*:

P. Gomes (✉)
Departmental Area of Mathematics, ISEL – Lisbon Superior Engineering Institute, Lisbon, Portugal
e-mail: pgomes@adm.isel.pt

N. Franco
CIMA-UE and Department of Mathematics, University of Évora, Évora, Portugal
e-mail: nmf@uevora.pt

L. Silva
CIMA-UE and Departmental Area of Mathematics, ISEL – Lisbon Superior Engineering Institute, Lisbon, Portugal
e-mail: lfs@adm.isel.pt

C. Grácio et al. (eds.), *Nonlinear Maps and their Applications*, Springer Proceedings in Mathematics & Statistics 57, DOI 10.1007/978-1-4614-9161-3_8,

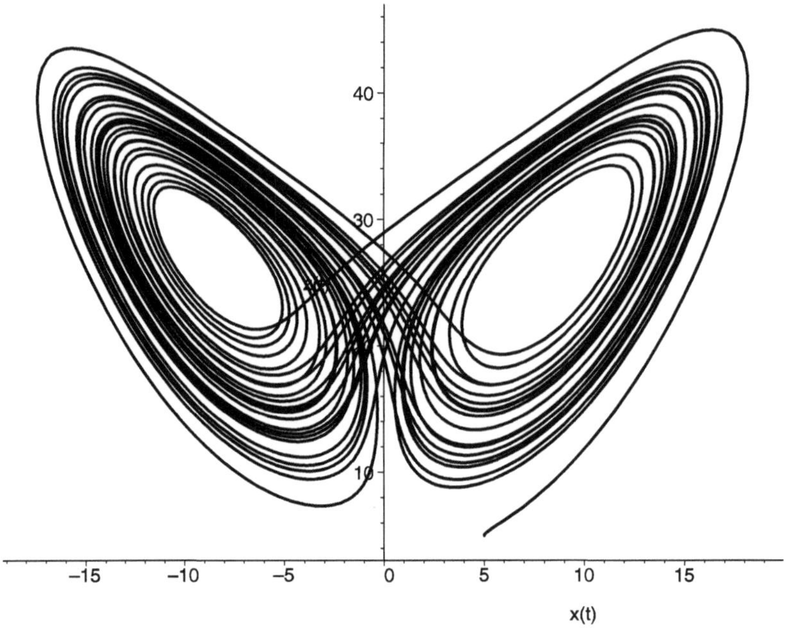

Fig. 8.1 The Lorenz attractor

$$x' = -10x + 10y, \quad y' = 28x - y - xz, \quad z' = -\frac{8}{3}z + xy. \tag{8.1}$$

The phase space of this system has a rich structure that provided one of the first examples of what came to be known as a *strange attractor*. Here, however, we are concerned only with its periodic orbits (Fig. 8.1).

8.2 Lorenz Knots and Links

Lorenz knots are the closed (periodic) orbits of the flow associated with the Lorenz equations. Lorenz links are collections of (possibly linked) Lorenz knots. A significant advance in the study of Lorenz knots and links was made by Williams [11, 12] when he introduced the *Lorenz template*, although a sketch of the idea had already appeared in Lorenz's original paper. The concept and its application to Lorenz knots and links was later developed by Birman and Williams, in [4], which marks the beginning of the systematic study of the subject. Birman has also recently published a review article on Lorenz knots [2].

8.2.1 The Lorenz Template

The Lorenz template, or knot-holder, is a branched 2-manifold built from two types of charts—joining and splitting—where an expanding semiflow is defined (Figs. 8.2 and 8.3). The arrows in Fig. 8.3 indicate the direction of the semiflow on the Lorenz template.

The Lorenz template has a remarkable feature: there is a bijection between the periodic orbits of the Lorenz flow in $\mathbf{R}^3$ and the periodic orbits of the semiflow on the template. This result, which came to be known as the *template theorem* is also due to Birman and Williams [4]:

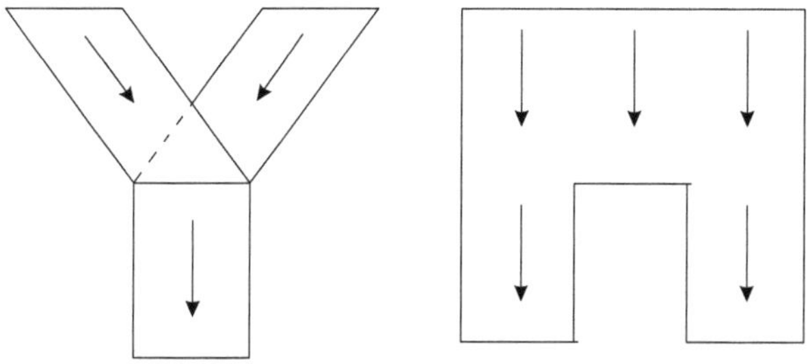

Fig. 8.2 Joining (*left*) and splitting (*right*) charts

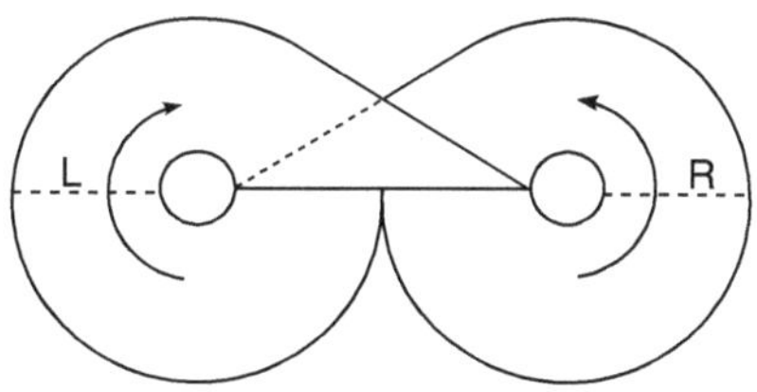

Fig. 8.3 The Lorenz template

Theorem 1. *Given a flow ϕ_t on a three-manifold M having a hyperbolic chain-recurrent set, the link of periodic orbits L_ϕ is in bijective correspondence with the link of periodic orbits L_τ on a particular embedded template $\tau \subset M$. On any finite sublink, this correspondence is via ambient isotopy.*

The study of Lorenz knots can thus be reduced to the study of the closed (periodic) orbits of the semiflow on the Lorenz template, which will also be referred to as Lorenz knots.

8.2.2 Lorenz Map

Each orbit on the Lorenz template crosses the horizontal line where the two branches of the joining chart meet—the *branch line* (Fig. 8.3). We can therefore define the first-return map on this line. The branch line can be mapped onto the interval $[-1,1]$, and the resulting *Lorenz map* f is a one-dimensional map of the interval $[-1,1]\setminus 0$ onto $[-1,1]$ with one discontinuity at the point $x=0$. The map is left undefined at $x=0$ and is strictly increasing in each subinterval of continuity (Fig. 8.4).

More precisely, a Lorenz map can be defined as follows [6]:

Definition 1. Let $P<0<Q$ and $r\geq 1$. A C^r Lorenz map $f:[P,Q]\to[P,Q]$ is a map described by a pair (f_-,f_+), where:

1. $f_-:[P,0]\to[P,Q]$ and $f_+:[0,Q]\to[P,Q]$ are continuous and strictly increasing maps.
2. $f(P)=P$, $f(Q)=Q$ and f has no other fixed points in $[P,Q]\setminus\{0\}$.
3. There exists $\rho>0$, the exponent of f, such that

$$f_-(x)=\tilde{f}_-(|x|^\rho) \text{ and } f_+(x)=\tilde{f}_+(|x|^\rho),$$

where $\tilde{f}_-$ and $\tilde{f}_+$, the coefficients of the Lorenz map, are C^r diffeomorphisms defined on appropriate intervals.

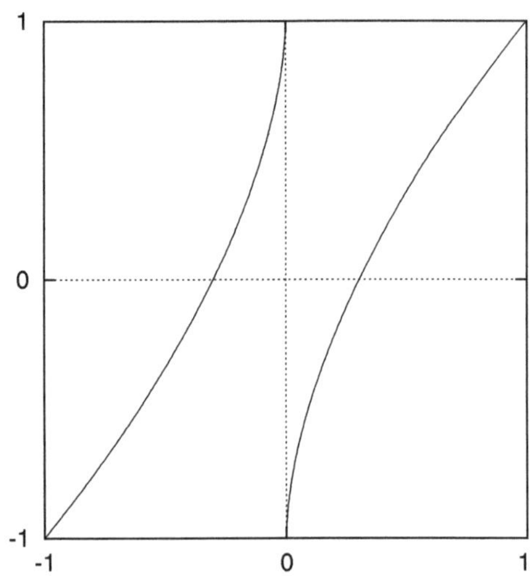

Fig. 8.4 Lorenz map

Each periodic orbit on the template determines a periodic orbit under the iteration of the Lorenz map, and conversely, each periodic orbit under the iteration of f uniquely determines a periodic orbit on the template under the semiflow. There is thus also a bijective correspondence between periodic orbits on the template and periodic orbits under the iteration of the Lorenz map.

8.3 Symbolic Dynamics

8.3.1 Symbolic Dynamics for the Lorenz Map

Let $f^j = f \circ f^{j-1}$ be the jth iterate of the Lorenz map f, and let f^0 denote the identity. We define the *itinerary* of a point x under the map as the symbolic sequence $(i_f(x))_j$, $j = 0, 1, \ldots$, where

$$(i_f(x))_j = \begin{cases} L \text{ if } f^j(x) < 0, \\ 0 \text{ if } f^j(x) = 0, \\ R \text{ if } f^j(x) > 0. \end{cases}$$

The itinerary of a point in $[-1,1] \setminus \{0\}$ under the Lorenz map can be either an infinite sequence in the symbols L, R or a finite sequence in L, R terminated by a single symbol 0 (because f is undefined at $x = 0$). The *length* $|X|$ of a finite sequence $X = X_0 \ldots X_{n-1}0$ is n, so it can be written as $X = X_0 \ldots X_{|X|-1}0$. The space Σ of all finite and infinite sequences can be ordered with the lexicographic order induced by $L < 0 < R$: given $X, Y \in \Sigma$, let k be the first index such that $X_k \neq Y_k$. Then $X < Y$ if $X_k < Y_k$ and $Y < X$ otherwise.

The *shift map* $s : \Sigma \setminus \{0\} \to \Sigma$ is defined as usual by $s(X_0 X_1 \ldots) = X_1 \ldots$ (it just deletes the first symbol). An infinite sequence X is periodic if there is $n > 0$ such that $s^n(X) = X$. A sequence X of (least) period n is denoted by $(X_0 \ldots X_{n-1})^\infty$, where $X_0 \ldots X_{n-1}$ is an aperiodic word in the symbols L, R.

A (finite or infinite) sequence X is called *L-maximal* if $X_0 = L$ and for $k > 0$, $X_k = L \Rightarrow s^k(X) \leq X$, and *R-minimal* if $X_0 = R$ and for $k > 0$, $X_k = R \Rightarrow X < s^k(X)$. An infinite periodic sequence $(X_0 \ldots X_{n-1})^\infty$ with least period n is L-maximal (resp. R-minimal) if and only if the finite sequence $X_0 \ldots X_{n-1}0$ is L-maximal (resp. R-minimal).

The *trip number t* is the number of syllables (subwords of type $L^a R^b$ with maximal length) in an aperiodic word. The trip number of a Lorenz link is the sum of the trip numbers of its components.

Each periodic orbit of the flow has a unique corresponding orbit in the Lorenz map, which in turn corresponds to the cyclic permutation class of one aperiodic word in the symbols L, R, since clearly, all periodic sequences resulting from shifting a given sequence represent points in the same periodic orbit (see [4]). Lorenz braids are thus also mapped to orbits in the symbolic space Σ, as Figs. 8.5 and 8.6 also illustrate.

8.3.2 The $*$-Product

A pair $(X, Y) \in \Sigma \times \Sigma$ is *admissible* if $X_0 = L$, $Y_0 = R$, and for $Z \in X, Y$ and all $k > 0$,

1. $Z_k = L \Rightarrow s^k(Z) \leq X$;
2. $Z_k = R \Rightarrow Y \leq s^k(Z)$.

Each of these inequalities is strict if, respectively, X or Y is finite. It can also be shown that a pair of periodic sequences $((X_0 \dots X_{n-1})^\infty,\ (Y_0 \dots Y_{m-1})^\infty)$ is admissible if and only if the associated pair of finite sequences $(X_0 \dots X_{n-1}0,\ Y_0 \dots Y_{m-1}0)$ is admissible.

These properties justify the usual identification between periodic maximal or minimal sequences $(X_0 \dots X_{n-1})^\infty$ with their finite counterparts $(X_0 \dots X_{n-1}0)$. Both can be identified by the finite aperiodic word $X_0 \dots X_{n-1}$, which we will do whenever appropriate.

The $*$-product between a pair of finite sequences (X,Y) and another finite sequence U is defined as

$$(X,Y) * U = \bar{U}_0 \dots \bar{U}_{|U|-1} 0,$$

where

$$\bar{U}_k = \begin{cases} X_0 \dots X_{|X|-1} & \text{if } U_k = L, \\ Y_0 \dots Y_{|Y|-1} & \text{if } U_k = R. \end{cases}$$

This identification between periodic and finite sequences allows us to extend the $*$-product to periodic sequences. Since both types of sequences can be identified with unique aperiodic words, we will also simply write the $*$-product of these words and interpret the result as a periodic symbolic sequence that represents a Lorenz knot.

8.3.3 Lorenz Braids

By cutting open the Lorenz template along the dotted lines under L and R in Fig. 8.3, we associate an open braid to every link on the template, which is called the *Lorenz braid* associated to the link. Every Lorenz link (that is, every set of closed orbits on the Lorenz template, including single knots) is the closure of a Lorenz braid, which in turn is uniquely determined by the link or knot (see [4]).

The strings in a Lorenz braid are divided into two sets: each string in the left set crosses over at least one of the strings on the right set. The rightmost left string always crosses over all the strings in the right set. There are no crossings between strings in the same group. Each Lorenz braid has an associated permutation—a *Lorenz permutation*—which completely determines the braid.

The Lorenz braid associated to a periodic symbolic sequence closes to a knot and can be obtained as follows:

1. Given an n-periodic sequence Z, write down its shifted sequences (the orbit of Z under the iteration of the shift map): $s(Z), s^2(Z), \dots, s^n(Z) = Z$. For example, if $Z = (LRRLR)^\infty$, then $s^1(Z) = (RRLRL)^\infty$, $s^2(Z) = (RLRLR)^\infty$, $s^3(Z) = (LRLRR)^\infty$, $s^4(Z) = (RLRRL)^\infty$, $s^5(Z) = (LRRLR)^\infty = Z$.
2. Order the shifted sequences lexicographically. In the same example, $s^3(Z) < s^5(Z) < s^2(Z) < s^4(Z) < s^1(Z)$.

3. If $s^k(Z)$ is in position i in the lexicographical order and $s^{k+1}(Z)$ is in position j, there is a string with start point i and end point j (Fig. 8.5).

A projection of the corresponding Lorenz knot can then be obtained by closing the braid (in this example, a trefoil knot) (Fig. 8.6).

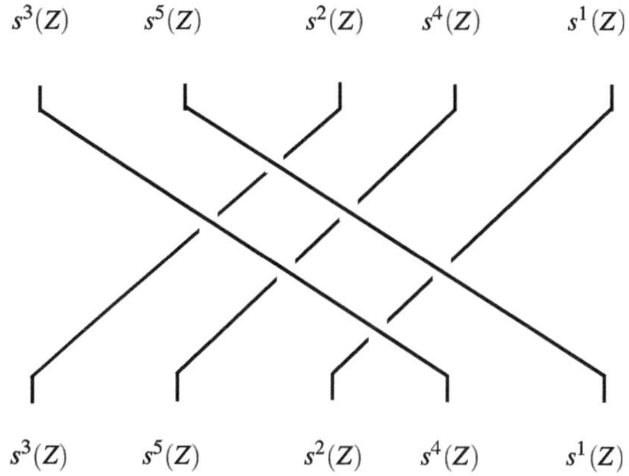

Fig. 8.5 Lorenz braid associated to $Z = (LRRLR)^\infty$

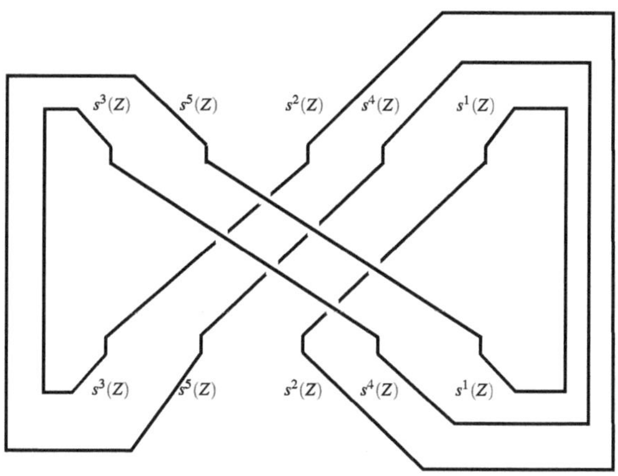

Fig. 8.6 Lorenz knot associated to $Z = (LRRLR)^\infty$

Invariants and Minimal Braids

There are several invariants associated to knots and links. One of them is the *braid index*, which is the minimum *string index* (the number of strings) among all closed braid representatives of the knot. Birman and Williams [4] first conjectured that the trip number (defined in Sect. 8.3.1) is the *braid index* of the link and therefore a knot invariant. This was later proved by Franks and Williams [7].

The *braid word* is the representation of a braid in n strings as a product of the generators $\sigma_1, \ldots, \sigma_{n-1}$ of the braid group B_n, where σ_i is the elementary braid in which string i crosses over string $i+1$. Birman and Williams also showed how to compute the braid word of this minimal braid, which is always a positive braid but not a Lorenz braid (a small typo in the original formula was corrected in [3]). Both Lorenz and minimal braids were used in our computations.

Remark 1. This completes a useful way to translate between periodic orbits on the Lorenz template, Lorenz braids, and aperiodic symbolic words (up to cyclic permutation). However, many (even an infinite number) orbits, braids, or words may represent the same knot, from a knot-theoretic point of view.

8.4 Hyperbolic Lorenz Knots

Complements of knots in S^3 are 3-manifolds. If such a manifold has a hyperbolic structure, the knot is called a *hyperbolic knot.*

A (p,q) *torus knot* (p,q relatively prime) is a knot that wraps around the standard solid torus T, p times in the longitudinal direction and q times in the meridional direction. A *satellite knot* is constructed as follows: Let K be a geometrically essential knot (intersecting every meridional disk) contained in a solid torus T. Let K_1 be another knot, and V a tubular neighborhood of K_1. If $h : T \to V$ is a homeomorphism, the knot $K_2 = h(K)$ is called a satellite of K_1, and K_1 is called a companion of K_2 [9].

Thurston [10] presented the following theorem, as a corollary to a more general result on hyperbolic structures of 3-manifolds:

Theorem 2. *If $K \subset S^3$ is a knot, then $S^3 - K$ has a geometric structure iff K is not a satellite knot. It has a hyperbolic structure iff in addition, K is not a torus knot.*

All knots (in particular, Lorenz knots) can thus be classified into three mutually disjoint classes: torus knots, satellites (which include composite knots), and hyperbolic knots. Complements of hyperbolic knots can be decomposed into ideal hyperbolic tetrahedra. The sum of the volumes of these tetrahedra is finite.

Much research has been done on hyperbolic knots. This includes estimates of the relative abundance of hyperbolic knots and the establishment of classes of hyperbolic knots [1]. In a recent article, Birman and Kofman [3] discuss the classification of Lorenz knots and provide a list of the 107 simplest hyperbolic knots that are also Lorenz knots. More precisely, their list includes the hyperbolic knots with a complement in S^3 that has seven or fewer ideal tetrahedra that are Lorenz knots. Five other knots from the same list were not definitely classified and may possibly also be Lorenz knots.

Birman and Williams [4] proved that all torus knots are Lorenz knots [4] and showed how to obtain them as the closure of corresponding Lorenz braids. Recently, Franco and Silva [6] constructed Lorenz braids associated to $*$-products of symbolic sequences. Interpreting results in [4] using this language, we conclude that all symbolic sequences that do not admit a $*$-product decomposition—*irreducible sequences*—are associated to torus knots. All symbolic sequences associated to hyperbolic knots must therefore be *reducible sequences* (sequences that admit a $*$-product decomposition).

8.4.1 Families of Hyperbolic Knots

Through examination of Birman and Kofman's list, we noticed that most of the knots listed can be obtained as members of families of $*$-products of symbolic sequences in Σ.

To test these sequences, we used the computer algebra system *Maxima* to generate symbolic sequences and braid words and then *SnapPy* to compute the hyperbolic volumes. *SnapPy* is a program for studying the topology and geometry of 3-manifolds developed and maintained by Marc Culler and Nathan Dunfield and based on the *SnapPea* kernel written by Jeff Weeks. *SnapPy* tries to find a hyperbolic structure for the manifold (in this case, the knot complement). If it cannot find one, it returns zero.

For each sequence, we computed the hyperbolic volume of the closure of the Lorenz braid and also of the closure of the Birman–Williams minimal braid defined above. Since both braids represent the same knot, the volumes computed should ideally be the same. This is not always the case, but it provides a way of checking the results.

Cao and Meyerhoff [5] proved in 2001 that the minimum volume for a knot complement is $2V = 2.02988\ldots$, where V is the volume of the ideal regular tetrahedron. This volume is the volume of the figure-eight knot, which is not a Lorenz knot. We therefore expected *SnapPy* to return, for hyperbolic knots, volumes strictly greater than this minimum and classified a knot as hyperbolic only when this was the case. The three families and the tests performed follow.

8.4.1.1 Families of $*$-Products

Family 0

The first family of sequences has the form

$$(X,Y)*Z = (LRL^k, RL^{k+2}(RL^{k+1})^p)*LRL^q,\ k,p>0,\ q\geq 0.$$

The string index of the corresponding Lorenz braid is $(p+q+2)(k+2)+1$, and the trip number (and braid index of the minimal braid) is $p+q+2$.

Knots $\mathbf{k}4_3$, $\mathbf{k}4_4$, $\mathbf{k}5_5$, $\mathbf{k}5_{14}$, $\mathbf{k}6_5$, $\mathbf{k}6_{11}$, $\mathbf{k}6_{14}$, $\mathbf{k}6_{17}$, $\mathbf{k}6_{25}$, $\mathbf{k}6_{27}$, $\mathbf{k}7_{27}$, $\mathbf{k}7_{30}$, $\mathbf{k}7_{34}$, $\mathbf{k}7_{37}$, and $\mathbf{k}7_{63}$ in Birman and Kofman's list belong to this family. We tested all sequences (19,530) with parameter values in the range $1 \leq k \leq 21$, $1 \leq p \leq 30$, $0 \leq q \leq 30$.

The computed volume was greater than the minimum value mentioned above for all sequences, so we conclude that all the corresponding knots are hyperbolic and present the following conjecture:

Conjecture 1. The Lorenz knots corresponding to the periodic symbolic sequences resulting from the products $(LRL^k, RL^{k+2}(RL^{k+1})^p) * LRL^q$, where $k > 0$, $p > 0$, and $q \geq 0$, are hyperbolic.

Family 1

The second family is composed of the sequences

$$(L(RL)^p, RL^k) * LR^n(LR)^m,\ p > 0,\ k > 1,\ n > 1,\ m \geq 0.$$

The string index of Lorenz braids is given by $1+2p+n(k+1)+m(k+2p+2)$, and the trip number is $p+n+(p+1)m$.

This family includes knots $\mathbf{k}3_1$, $\mathbf{k}5_1$, $\mathbf{k}5_6$, $\mathbf{k}5_{11}$, $\mathbf{k}7_8$, $\mathbf{k}7_{13}$, $\mathbf{k}7_{28}$, $\mathbf{k}7_{47}$, $\mathbf{k}7_{58}$, and $\mathbf{k}7_{102}$ in the above-mentioned list. We tested all 5,600 sequences with $1 \leq p \leq 7$, $2 \leq k \leq 11$, $2 \leq n \leq 11$, and $0 \leq m \leq 7$. *SnapPy* returned a volume above the minimum for all sequences except those with $p = 1$ and $m = 0$ and the single sequence $LRLRLRLLRLL = (LRLRL, RLL) * LRR$ corresponding to $p = 2$, $k = 2$, $n = 2$, and $m = 0$.

We then tested an additional 3,000 pseudorandom sequences (using Maxima's "random" command) with $1 \leq p \leq 7$, $2 \leq k \leq 20$, $2 \leq n \leq 20$, and $0 \leq m \leq 7$. All the nonhyperbolic cases were included in the mentioned exception. These results led us to our second conjecture:

Conjecture 2. The Lorenz knots corresponding to the periodic symbolic sequences resulting from the products $(L(RL)^p, RL^k) * LR^n(LR)^m$, $p > 0$, $k > 1$, $n > 1$, $m \geq 0$ are hyperbolic except if $p = 1$ and $m = 0$ or $p = 2$, $k = 2$, $n = 2$, and $m = 0$.

Family 2

The third family has the form

$$(LRLL, RL^{n_1} \ldots RL^{n_k}) * Z,\ |Z| > 1, n_1 > \ldots n_k > 2.$$

Here Z is an arbitrary word, so this family is in a sense more general than the preceding two. The string index and trip number of the product depend on the word Z, and a formula for both involves the number of L and R symbols in Z and the number of exponents n_i. Knots $\mathbf{k}4_3$, $\mathbf{k}7_{37}$, $\mathbf{k}7_{82}$, and $\mathbf{k}7_{90}$ are in this family. The string index of the Lorenz braid is $4n_L + (k + n_1 + \cdots + n_k)n_R$, where n_L (n_R) is the number of L (R) symbols in Z and k is the number of exponents n_i. The trip number is given by $n_L + kn_R$. For this family, 19,929 products were tested, with all the possible combinations of exponents such that $12 > n_1 > \ldots n_k > 2$ and $1 < |Z| < 8$.

Only 69 of these sequences had zero volume; they were all found to be products in which $Y = RL^{n_1} \ldots RL^{n_k}$ has trip number 1 (there is only one exponent n_1). However, there are also many hyperbolic knots for which Y has trip number 1. This led us to consider a subfamily satisfying the above condition and formulate our last conjecture:

Conjecture 3. The Lorenz knots corresponding to the products $(LRLL, Y) * Z$, where $Y = RL^{n_1} \ldots RL^{n_k}$, $|Y| > 1$, $|Z| > 1$, are hyperbolic.

Hardware and Software

Hardware

Laptop INSYS

- CPU: Intel Core i7 CPU M620 @2.67GHz
- Memory (RAM): 8GB

Software

- OS: Debian GNU/Linux 6.0.4 "Squeeze" (`http://www.debian.org/`)
- Linux kernel 2.6.32-5-amd64
- Maxima 5.21.1 (`http://maxima.sourceforge.net/`)
- SnapPy 1.5.1 (`snappy.computop.org`)
- Python 2.6.6 (`http://www.python.org/`)

References

1. Adams, C.: Hyperbolic knots. In: Menasco, W., Thistlethwaite, M. (eds.) Handbook of Knot Theory, pp. 1–18. Elsevier, Amsterdam (2005)
2. Birman, J.: Lorenz knots. arXiv:1201.0214 [math.GT] `http://arxiv.org/abs/1201.0214`
3. Birman, J., Kofman, I.: A new twist on Lorenz links. J. Topol. **2**, 227–248 (2009)
4. Birman, J., Williams, R.: Knotted periodic orbits in dynamical systems–I: Lorenz's equations. Topology **22**, 47–82 (1983)
5. Cao, C., Meyerhoff, G.: The orientable cusped hyperbolic 3-manifolds of minimum volume. Invent. Math. **146**, 451–478 (2001)
6. Franco, N., Silva, L.: Genus and braid index associated to sequences of renormalizable Lorenz maps. Discret. Contin. Dyn. Syst. **32**, 565–586 (2012)
7. Franks, J., Williams, R.: Braids and the Jones polynomial. Trans. Amer. Math. Soc. **303**, 97–108 (1987)

8. Lorenz, E.: Deterministic non-periodic flow. J. Atmos. Sci. **20**, 130–141 (1963)
9. Rolfsen, D.: Knots and Links. AMS Chelsea Publishing, Providence (2003)
10. Thurston, E.: Three-dimensional manifolds, Kleinian groups and hyperbolic geometry. Bull. Amer. Math. Soc. (N.S.) **6**, 357–381 (1982)
11. Williams, R.: The structure of Lorenz attractors. In: Bernard, P., Ratiu, T. (eds.) Turbulence Seminar (Berkeley 1976/77). Lecture Notes in Mathematics, vol. 615, pp. 94–112. Springer, Berlin (1977)
12. Williams, R.: The structure of Lorenz attractors. Publ. Math. I.H.E.S. **50**, 73–99 (1979)

Chapter 9
Cycle Statistics in Complex Networks and Ihara's Zeta Function

Clara Grácio, Anthony C.C. Coolen, and Alessia Annibale

Abstract Network representations are popular tools for characterizing and visualizing patterns of interaction between the microconstituents of large, complex synthetic, social, or biological systems. They reduce the full complexity of such systems to topological properties of their associated graphs, which are more amenable to analysis. In particular, the cyclic structure of complex networks is receiving increasing attention, since the presence of cycles affects strongly the behavior of processes supported by these networks. In this paper, we survey the analysis of cyclic properties of networks, and in particular the use of Ihara's zeta function for counting cycles in networks.

9.1 Introduction

In recent years, many complex systems in physics, biology, computer science, economics, and the social sciences have been studied in terms of networks that represent the interaction patterns between interacting variables [1–3]. The structural characterization of complex networks is an important problem, since from an understanding of the geometric and topological properties of interactions we can learn much about the emerging behavior and global order of the processes for which networks serve as infrastructures.

C. Grácio (✉)
Department of Mathematics, University of Evora, Evora, Portugal,
CIMA-UE, Evora, Portugal
e-mail: mgracio@uevora.pt

A.C.C. Coolen • A. Annibale
Kings College London, London, UK
e-mail: ton.coolen@kcl.ac.uk; alessia.annibale@kcl.ac.uk

C. Grácio et al. (eds.), *Nonlinear Maps and their Applications*, Springer Proceedings in Mathematics & Statistics 57, DOI 10.1007/978-1-4614-9161-3_9,

The cyclic structure of networks is of particular interest, since the presence of cycles is known to affect significantly the dynamic properties and functional robustness of many different processes, including traffic congestion, shipping processes, and epidemic spreading. Also, the distinct dynamical and thermodynamic properties of spin processes on large low-dimensional lattices (special cases of regular graphs), compared to processes on *random* graphs (even if these have identical coordination numbers as those of the lattices), are predominantly due to the high density of short cycles in lattices.

In biology, proteomic and gene regulatory networks are the infrastructure of cellular signaling. Errors in cellular signaling are responsible for many diseases, such as cancer, autoimmune illnesses, and diabetes. Therefore, understanding how the structure of signaling networks affects the flow of information is a very important question. Also, here short cycles are the main local network modules, so any realistic model for signaling networks should incorporate cycles as key observables. Networks with short loops are quite difficult to handle, hence most modeling in this field has so far been limited to locally treelike graphs with controlled degree statistics and correlations. The next step would be to define and analyze random graph ensembles in which also the number of short cycles is controlled. However, evaluating the number of cycles of a given length for an arbitrary graph is a nontrivial problem. Indeed, many studies focus instead on closed paths, whose number follows for every graph from the trace of powers of the connectivity matrix, or equivalently from the spectral density, which can be computed at a much lower computational cost.

In 1960, Yasutaka Ihara [4] introduced a function that today is known as the Ihara zeta function, which takes the form of an infinite product of a set of equivalence classes of paths in the graph. Later, Bass [5] proved that this function is a rational polynomial, which is determined by the graph in question, through the adjacency matrix, the degree of each vertex, and the size of the sets of vertices and edges of the graph. In 1996, Stark and Terras [6] introduced the edge zeta function of a finite graph as a generalization of the Ihara zeta function. Our goal here is to show that the edge zeta function, which can be computed in polynomial time, can be used to calculate the cycle structure of an arbitrary graph in polynomial time. It is therefore a powerful and promising practical tool with which to calculate important information about a graph. Application of this formalism to protein interaction networks is currently in progress.

9.2 Network and Graph Preliminaries

Finite graphs. A finite graph is a set $G = (V(G), E(G))$, where $V(G)$ is a set of $N > 0$ vertices or nodes (labeled by $i = 1 \dots N$) and $E(G)$ is the set of $M \geq 0$ edges or links e_{ij} that connect vertices i and j [7]. The edge set is unordered for undirected graphs and ordered for directed graphs. A weighted graph is one in which with each pair of vertices (i, j) there is associated a nonnegative weight $a_{ij} \in \mathbb{R}$ such that $a_{ij} = 0$ if the vertices i and j are not connected. If the graph is not weighted, then $a_{ij} = 1$ if i

and j are connected, and $a_{ij} = 0$ otherwise. If the graph is not directed, $a_{ij} = a_{ji}$ for al (i, j). The matrix $A(G) = [a_{ij}]$, with $i, j \in V(G)$, is called the adjacency matrix of the graph G. A graph is undirected if and only if its adjacency matrix is symmetric (the proof is trivial). A (self-)loop is an edge of the form e_{ij} with $i = j$. A graph is called *simple* if it contains no loops and if no ordered pair of vertices appears more than once in E. In this paper we will consider only simple finite unweighted graphs.

Local topology characterization. The neighborhood $\partial_i = \{j \in V(G) : e_{ij} \in E(G)\}$ of a vertex i in a simple undirected graph G is the set of vertices that have a link to i. It can equivalently be written as $\partial_i = \{j \in V(G) : a_{ij} = 1\}$. To characterize the topology of G in the vicinity of node i, one can define [8] the degree $k_i = |\partial_i|$, i.e., the number of links to node i, and the clustering coefficient $c_i \in [0, 1]$ (defined only for nodes with $k_i > 1$), which measures how densely connected the neighbors of node i are. These can both be written in terms of the adjacency matrix $A(G) = [a_{ij}]$:

$$k_i = \sum_{i=1}^{N} a_{ij}, \tag{9.1}$$

$$c_i = \frac{\sum_{j,m=1}^{N}(1-\delta_{jm})a_{ij}a_{jm}a_{mi}}{\sum_{j,m=1}^{N}(1-\delta_{jm})a_{ij}a_{mi}} = \frac{\sum_{j,m=1}^{N}(1-\delta_{jm})a_{ij}a_{jm}a_{mi}}{k_i(k_i-1)}, \tag{9.2}$$

with the Kronecker symbol $\delta_{ij} = 1$ if $i = j$, and $\delta_{ij} = 0$ if $i \neq j$. The maximum degree of graph G is denoted by $\Delta(G)$, and the minimum degree by $\delta(G)$. For directed graphs, the above definitions are generalized in a simple way, by distinguishing between in-degrees $k_i^{\text{in}} = \sum_j a_{ij}$ (the number of links into node i) and out-degrees $k_i^{\text{out}} = \sum_j a_{ji}$ (the number of links away from i).

Global topology characterization. One can characterize a graph's topology at a global level by giving averages or distributions of the local site-specific observables. For instance, the average degree would be $\langle k \rangle = N^{-1} \sum_i k_i$, and the degree distribution $P(k)$ in an undirected graph (giving the probability that a randomly selected node has exactly k edges) is $P(k) = N^{-1} \sum_{i=1}^{N} \delta_{k,k_i}$. Similarly, the average clustering coefficient [9], calculated over those nodes i for which c_i is defined, is

$$c = \frac{\sum_{i,k_i>1} c_i}{\sum_{i,k_i>1} 1}. \tag{9.3}$$

Global degree–degree correlations in undirected graphs would be captured by the joint degree distribution of connected nodes [10]:

$$W(k, k') = \frac{\sum_{ij} a_{ij} \delta_{k,k_i} \delta_{k,k_j}}{\sum_{ij} a_{ij}} = \frac{1}{N\langle k \rangle} \sum_{ij} a_{ij} \delta_{k,k_i} \delta_{k,k_j}. \tag{9.4}$$

Again, generalization of such distributions to directed graphs is straightforward; see, e.g., [11]. For graphs known to have a modular structure, there are also specific measures for characterizing the extent of modularity [3].

Closed walks and cycles in graphs. A sequence of vertices $\{i_0,\ldots,i_k\}$ in a graph G will be called a k-walk, with initial vertex i_0 and terminal vertex i_k, if for each index $j\in\{0,\ldots,k-1\}$, there exists an edge $e_{i_j,i_{j+1}}\in E(G)$, or equivalently $(\forall j\in\{0,\ldots,k-1\}):\ a_{i_{j+1},i_j}=1$. A k-walk thus contains k edges. A closed k-walk is one in which $i_0=i_k$. A k-cycle is a closed non-self-intersecting k-walk, i.e., a closed k-walk in which no vertex appears more than once, with the exception of $i_0=i_k$. We define two closed k-walks to be equivalent if they can be mapped into each other by repeated operation of sequence shift operations

$$\{i_0,i_1,\ldots,i_{k-1},i_0\}\to\{i_1,\ldots,i_k,i_0,i_1\}$$

or sequence reversals

$$\{i_0,i_1,\ldots,i_{k-2},i_{k-1},i_0\}\to\{i_0,i_{k-1},i_{k-2},\ldots,i_1,i_0\}.$$

The number of short closed walks and short cycles in a graph has a significant impact on all collective processes that have this graph as their interaction representation. In the remainder of this paper, we will be concerned with the calculation of the number of k-walks and k-cycles in graphs.

9.3 Counting k-Walks

Relation to the eigenvalue spectrum of the adjacency matrix. We review some useful results on the number of walks in graphs. These show that the number of inequivalent closed k-walks in an undirected graph follows from the spectrum of its adjacency matrix.

Lemma 1. *Let G be a (directed or undirected) graph with associated adjacency matrix A. The number of k-walks from node i to node j in G is the (i,j)th entry of the matrix A^k.*

Proof (By induction on k). The claim is true for $k=1$, since $A^1=A$. Suppose the claim is true for $k=\ell$. Since each $\ell+1$-walk from node j to node i can be constructed as an ℓ-walk from j to an arbitrary node w with $a_{iw}=1$, followed by a final step from node w to node i, we trivially obtain

$$\text{number of } k\text{-walks from } j \text{ to } i=\sum_{w=1}^{N}a_{iw}(A^{\ell})_{wj}=(A^{\ell+1})_{ij}.$$

Hence the claim is true also for $\ell+1$. The result follows by induction. □

Lemma 2. *Let G be an undirected graph with associated adjacency matrix A, and let $\{\lambda_1, \ldots, \lambda_N\}$ be the sequence of N eigenvalues of A. The number $\mathcal{L}_k(G)$ of non-equivalent closed k-walks in G is given by $\mathcal{L}_k(G) = \frac{1}{2k} \sum_{i=1}^{N} \lambda_i^k$.*

Proof. It follows from Lemma 1 that the number of closed k-walks starting from node i equals $(A^k)_{ii}$. The total number of closed k-walks is therefore $\mathrm{tr}(A^k) = \sum_i (A^k)_{ii}$. Since the trace of a symmetric matrix is the sum of its eigenvalues (which are all real), and since each eigenvalue λ_i of A corresponds to an eigenvalue λ^k of A^k, the total number of closed k-walks is $\sum_i \lambda_i^k$. Finally, there are k distinct shift operations that map each k-walk to a distinct but equivalent k-walk, and one reversal. Therefore, each individual k-walk is a member of a set containing $2k$ mutually equivalent members. Hence the claim. □

Simple spectral properties of symmetric adjacency matrices. In view of Lemma 2, we will be interested in spectral properties of the adjacency matrices of undirected graphs G. Since each such adjacency matrix is symmetric, there are always N eigenvalues, where N is the number of nodes in G, and all eigenvalues are real. Let us briefly recall two useful bounds on the largest eigenvalue $\lambda_{\max}$ of the adjacency matrix of an undirected graph:

Lemma 3. *Let $\lambda_{\max}$ be the largest eigenvalue of the adjacency matrix A of an undirected graph G with degree sequence $\{k_1, \ldots, k_N\}$. Let the maximum degree and the average degree of G be $\Delta(G) = \max_{i \le i \le N} k_i$ and $\langle k \rangle = N^{-1} \sum_{i=1}^{N} k_i$. Then the following inequalities hold:*

$$\langle k \rangle \le \lambda_{\max} \le \Delta(G).$$

Proof. Since the matrix A is symmetric, the inequality $\langle k \rangle \le \lambda_{\max}$ follows directly from the Rayleigh–Ritz theorem (we abbreviate $\mathbf{y} = (y_1, \ldots, y_N)$):

$$\begin{aligned}
\lambda_{\max} &= \max_{\mathbf{y} \in \mathbb{R}^N} \frac{\mathbf{y} \cdot A\mathbf{y}}{\mathbf{y}^2} = \max_{\mathbf{y} \in \mathbb{R}^N} \frac{\sum_{i,j=1}^{N} y_i a_{ij} y_j}{\sum_{i=1}^{N} y_i^2} \ge \left. \frac{\sum_{i,j=1}^{N} y_i a_{ij} y_j}{\sum_{i=1}^{N} y_i^2} \right|_{y_k = 1\ \forall k} \\
&= \frac{1}{N} \sum_{i,j=1}^{N} a_{ij} = \frac{1}{N} \sum_{i=1}^{N} k_i = \langle k \rangle.
\end{aligned}$$

The proof of the second inequality follows that of the Perron–Frobenius theorem. Since A is symmetric, it has N eigenvectors that can be chosen real-valued. Let $\mathbf{y} \in \mathbb{R}^N \neq \mathbf{0}$ be an eigenvector of A with eigenvalue $\lambda_{\max}$. If all its components are non-positive, we replace $\mathbf{y}$ by $-\mathbf{y}$. Now there is at least one index i with $y_i > 0$; we identify the largest of the positive components, $y_\ell = \max_{i, y_i > 0}$. They may be more than one such index ℓ. Since $\mathbf{y}$ is an eigenvector of A with eigenvalue $\lambda_{\max}$, and $a_{\ell j} y_j \le a_{\ell j} y_\ell$ for all j, we write

$$0 = \lambda_{\max} y_\ell - \sum_{j \neq \ell}^{N} a_{\ell j} y_j \ge \lambda_{\max} y_\ell - \sum_{j \neq k}^{N} a_{\ell j} y_\ell = (\lambda_{\max} - k_\ell) y_\ell.$$

With $(\lambda_{\max} - k_\ell)y_\ell \leq 0$ having been established, it follows from $y_\ell > 0$ that $\lambda_{\max} \leq k_\ell$. Hence also $\lambda_{\max} \leq \max_{\ell=1\ldots N} k_\ell = \Delta(G)$. This completes the proof. □

In particular, it now follows directly as a simple corollary that the number of non-equivalent closed walks in a simple undirected graph G is bounded by

$$\mathcal{L}_k(G) \leq \frac{N}{2k}\Delta^k(G). \tag{9.5}$$

This and other similar inequalities can be obtained in many alternative ways.

Counting 3-cycles. Due to the fact that closed walks with $k = 3$ cannot be self-intersecting, the number $\mathcal{L}_3(G)$ of non-equivalent closed 3-walks equals the number $\mathcal{C}_3(G)$ of 3-cycles (usually called "triangles") in any simple finite graph G. Hence

$$\mathcal{C}_3(G) = \mathcal{L}_3(G) = \frac{1}{6}\operatorname{tr}(A^3). \tag{9.6}$$

For undirected graphs, we saw that the trace can also be written in terms of the eigenvalue spectrum of A; for directed graphs, this is generally not possible. The traces of lower powers of the adjacency matrix A also have simple interpretations. For instance, due to the absence of self-loops, every finite simple graph has $\operatorname{Tr}(A) = \sum_{i=1}^{N} a_{ii} = 0$, and for undirected graphs, we may write (using $a_{ij}a_{ji} = a_{ij}^2 = a_{ij}$),

$$\operatorname{Tr}(A^2) = \sum_{i,j=1}^{N} a_{ij}a_{ji} = \sum_{i,j=1}^{N} a_{ij} = \sum_{i=1}^{N} k_i = N\langle k\rangle. \tag{9.7}$$

9.4 Counting k-Cycles for $k > 3$

The equivalence between closed walks and cycles for $k = 3$ does not generalize to lengths $k \geq 4$, since as soon as $k > 3$, it is possible for a k-walk to self-intersect or to exhibit backtracking; see, e.g., Fig. 9.1 for an explicit example to illustrate this. To calculate the cycle numbers $\mathcal{C}_{k>3}(G)$, we need an alternative method; this will be based on zeta functions for graphs.

Zeta functions—an introduction. In attempting to find a formula for calculating the number of primes smaller than a given number n, Georg Friedrich Bernhard Riemann investigated the series

$$\zeta(z) = \sum_{n=1}^{\infty} \frac{1}{n^z}, \qquad z(= x + iy) \in \mathbb{C}, \tag{9.8}$$

which today is known as the Riemann zeta function. The zeta function ζ can be continued analytically to a meromorphic function in $\mathbb{C}$, with a simple pole at $z = 1$. The residue at the pole is 1. There is a different way to express the zeta function that illustrates how it connects to the prime numbers. Riemann's article [12] quotes a result of Euler (the Euler product):

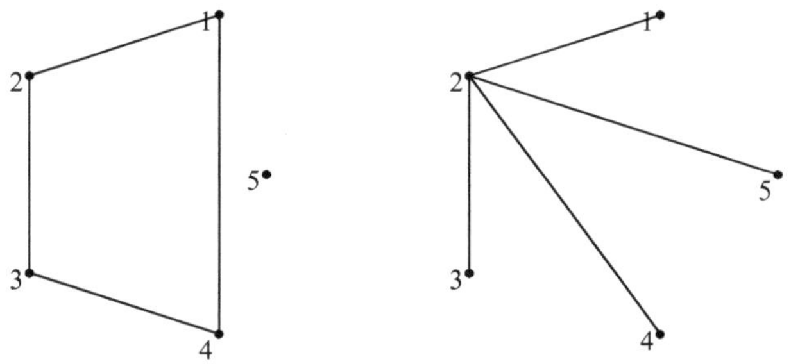

Fig. 9.1 Example of two undirected graphs with $N = 5$ nodes. The one on the *left* has a 4-cycle, whereas the graph on the *right* has no cycles. Yet both have the same spectrum, and hence the same numbers $\mathcal{L}_k(G)$ of closed walks of any length. The eigenvalue polynomial of both adjacency matrices is $P(\lambda) = \lambda^5 - 4\lambda^3$, so for both graphs, $\mathcal{L}_k(G) = 0$ for odd k and $\mathcal{L}_k(G) = 2^k/k$ for even k. However, in the graph on the *right*, all closed walks involve backtracking, whereas on the *left*, there is one closed 4-walk that is a true cycle

$$\zeta(z) = \prod_{p \text{ prime}} \left(1 - p^{-z}\right)^{-1}. \tag{9.9}$$

This connects the distribution of primes to the roots of the zeta function. Riemann noted that the even negative integers $-2, -4, -6, \ldots$ are zeros of the function, called trivial zeros, and that any nontrivial zero lies in the open strip $\{s \in \mathbb{C} : 0 < Re(s) < 1\}$, which is called the critical strip. Riemann then presented his famous conjecture:

Conjecture 1 (Riemann hypothesis). The nonreal zeros of $\zeta(z)$, i.e., those with $0 < Re(z) < 1$, are on the "critical line" $Re(z) = \frac{1}{2}$.

More than one hundred years later, this conjecture is still open. It is undoubtedly one of the most important problems in mathematics.

One can define alternative zeta functions whereby the role of prime numbers is represented by other mathematical objects. For instance, If one considers a compact Riemann surface S of genus 2 endowed with a metric of constant curvature -1, this surface can be represented by a quotient space H^2/Γ of the upper half-plane H^2 of a Fuchsian group Γ that is isomorphic to a fundamental group of $\mathcal{M}$. With primes defined to be the geodesics on S, the Selberg trace formula [13, 18], which has been of great interest to mathematicians for almost 50 years, shows that there is a relation between the length spectrum of these prime geodesics and the spectrum of the Laplace operator on S:

$$\begin{aligned} \mathrm{Tr}(e^{t\Delta_{\mathcal{M}}}) = \sum e^{-t\lambda_k} = area(\mathcal{M}) \frac{e^{-\frac{t}{4}}}{(4\pi t)^{\frac{3}{2}}} \int_0^{+\infty} \frac{b e^{-\frac{b^2}{4t}}}{\sinh(\frac{1}{2}b)} db \\ + \frac{1}{2} \sum_{n=1}^{\infty} \sum_{\ell(p)} \frac{\ell(p)}{\sinh(\frac{1}{2}\ell(p^n))} \frac{e^{-\frac{t}{4}}}{(4\pi t)} e^{\frac{-(\ell(p^n))^2}{4t}}. \end{aligned} \tag{9.10}$$

The trace formulas, as is presently understood, are a general identity

$$\sum\{\text{geometric terms}\} = \sum\{\text{spectral terms}\}$$

In this case, the Selberg trace formula is formally similar to the explicit formulas relating the zeros of the Riemann zeta function to prime numbers, with the zeta zeros corresponding to eigenvalues of the Laplacian. For this work, Selberg introduced a function, which was later called the Selberg zeta function associated to a compact Riemannian manifold, whose analytic properties are encoded by the Selberg trace formula:

$$Z(s) = \prod_{[p]}\prod_{j\geq 0}\left(1 - e^{-(s+j)\ell(p)}\right), \tag{9.11}$$

where p are primitive closed geodesics (primitive means they go around only once) and $\ell(p)$ is the length of the geodesic p.

The Ihara zeta function. In this paper, we will be interested in applications of the zeta function and Euler's expansion to graph theory, by identifying cycles in a graph as "primes." In the 1960s, Ihara [4, 14] defined the Ihara zeta function for finite k-regular graphs. Since then, much work has been done to remove the restrictive condition of graph regularity and to simplify the factorization of the zeta function. Due to its many applications, there has been an increase recently in the study of Ihara's zeta function. We will utilize the Euler product expansion of the zeta function for the purpose of counting cycles.

In Ihara's formalism, a graph may have multiple edges, but here we limit ourselves to simple graphs. A cycle is called primitive if it is not a power of another cycle. We defined earlier the concept of equivalent closed walks and cycles; for undirected graphs, they are related to each other by order-preserving vertex shifts and order reversal of the vertex sequence; for directed graphs, they are related to each other by order-preserving vertex shifts only. We denote by $[c]$ the class of all cycles equivalent to a given cycle c, and we let $\mathbb{P}(G)$ be the set of all prime cycle classes of a finite connected graph G. We denote by $\ell([c])$ the number of edges in a representative cycle c of equivalence class $[c]$.

Definition 1. The Ihara zeta function of a simple finite graph G, for sufficiently small $u \in \mathbb{C}$, is given by

$$Z_G(u) = \prod_{[c]\in\mathbb{P}(G)}\left(1 - u^{\ell([c])}\right)^{-1}. \tag{9.12}$$

For any simple finite graph G with N nodes and M edges, the reciprocal $1/Z_G(u)$ is a finite polynomial in u, of order $\sum_{[c]\in\mathbb{P}(G)}\ell([c])$. In fact, it turns out that the order of this polynomial cannot exceed $2M$, so

$$1/Z_G(u) = c_0 + c_1u + c_2u^2 + c_3u^3 + \cdots + c_{2m}u^{2M}. \tag{9.13}$$

The coefficients c_i $(i = 1,\ldots,2M)$ give us information about the cyclic structure of the graph. They are called the coefficients of the Ihara zeta function. Similar to the ordinary zeta function, an alternative expression exists for Ihara's zeta function that allows us to calculate $Z_G(u)$ without invoking "primes" (i.e., cycles) explicitly. Let the $N \times N$ diagonal matrix D be defined by the entries $d_{ij} = k_i \delta_{ij}$, where k_i is the degree of the vertex i, and define the $N \times N$ matrix $Q = D - I$ (where I is the $N \times N$ identity matrix).

Theorem 1. *The Ihara zeta function of a simple finite graph G with $|V(G)| = N$ vertices and $|E(G)| = M$ edges can be written as*

$$Z_G(u) = \frac{\left(1-u^2\right)^{N-M}}{\det\left(I - Au + Qu^2\right)}. \tag{9.14}$$

Proof. See [5,6,15]. □

In 2000, Kotani and Sunada [16] gave an explicit expression for c_{2M}:

Theorem 2. *Let G a finite graph with $|V(G)| = N$ vertices and $|E(G)| = M$ edges, and let $Z_G(u)$ denote its Ihara zeta function. Let k_i be the degree of a node i (the number of edges incident on it). Then*

$$c_{2M} = (-1)^{M-N} \prod_{i=1}^{N} (k_i - 1). \tag{9.15}$$

Similarly, coefficients of small powers in the expansion of $1/Z_G(u)$ are calculated relatively easily; for instance, since by definition, $\ell([c]) > 0$ for all $[c]$, one can immediately read off from the definition of the coefficients that

$$c_1 = - \sum_{[c] \in \mathbb{P}(G)} \delta_{\ell([c]),1}, \tag{9.16}$$

which is zero if G is simple (here G has no loops, so $\ell([c]) > 0$ for all c). Many other properties relating to the coefficients c_i have been established over the years; see, e.g., [17].

The Ihara–Selberg zeta function. There is a natural correspondence between the set of symmetric simple graphs and the set of all graphs, obtained by identifying edges of a graph G with vertices in its associated edge graph $D(G)$. The construction is as follows. One first represents each undirected edge in G in the usual way by two nonzero entries $a_{ij} = a_{ji} = 1$ of the adjacency matrix $A(G)$, i.e., by two distinct nonzero directed edges that connect the nodes i and j. Each pair (i, j) with $a_{ij} = 1$ is defined to be a vertex of the edge graph $D(G)$, and two such vertices $(i, j) \in V(D(G))$ and $(r,s) \in V(D(G))$ are connected by a directed edge in $E(D(G))$ if the G-edge (r,s) "feeds into" the G-edge (i, j) and these two edges are not connecting the same two vertices, i.e., if $a_{rs} = a_{ij} = 1$, $r = j$, and $s \neq i$. So the edge graph $D(G)$ has the $2|E(G)| \times 2|E(G)|$ adjacency matrix T, defined as

$$T_{(i,j),(r,s)} = a_{ij}a_{rs}\delta_{rj}(1-\delta_{si}). \tag{9.17}$$

There is a one-to-one correspondence between primitive cycles with no backtracking or tail in G and primitive cycles in $D(G)$. Moreover, if G is a connected graph, the zeta function $Z_G(u)$ can be written in terms of the adjacency matrix T of the edge graph $D(G)$ as [6]:

$$Z_G(u) = \det(I-Tu)^{-1}. \tag{9.18}$$

Combining this result with (9.13) tells us that

$$\begin{aligned}\det(I-Tu) &= c_0 + c_1 u + c_2 u^2 + c_3 u^3 + \cdots + c_{2m}u^{2m} \\ &= \prod_{[c]\in\mathbb{P}(G)} \left(1-u^{\ell([c])}\right).\end{aligned} \tag{9.19}$$

Thus we can calculate the coefficients of the zeta function, from which in turn we can extract valuable information on the cyclic structure of G, from the eigenvalue polynomial of its associated edge graph $D(G)$. We can write the polynomial representation of $1/Z_G(u)$ also explicitly in terms of the cycle numbers $\mathcal{C}_\ell(G)$, via

$$\begin{aligned}\prod_{[c]\in\mathbb{P}(G)} \left(1-u^{\ell([c])}\right) &= \prod_{\ell\geq 2}(1-u^{\ell})^{\mathcal{C}_\ell(G)} \\ &= \prod_{\ell\geq 2}\left\{\sum_{n=0}^{\mathcal{C}_\ell(G)} \binom{\mathcal{C}_\ell(G)}{n}(-1)^n u^{\ell n}\right\}.\end{aligned} \tag{9.20}$$

Working out the lowest orders in u then gives us the following result.

Proposition 1. *Let G be a simple undirected graph with Ihara zeta function $Z_G(u)$.*

(1) The coefficient c_1 is zero due to G being simple.
(2) The coefficient c_2 is minus twice the number of primitive cycles of length 2 in G, and the coefficients c_3, c_4, and c_5 are respectively minus twice the number of triangles, squares, and pentagons in G.
(3) The coefficient c_6 is minus twice the number of hexagons in G plus four times the number of pairs of edge-disjoint triangles plus twice the number of pairs of triangles with a common edge.
(4) The coefficient c_7 is the negative of twice the number of heptagons in G plus four times the number of edge-disjoint pairs of one triangle and one square plus twice the number of pairs of one triangle and one square that share a common edge.

Explicit application of Ihara's zeta function. In the remainder of this paper we illustrate the application of Ihara's zeta function to the counting of cycles, by working out one specific example in full detail: the $N=4$ graph G shown in Fig. 9.2. To construct the associate edge graph $D(G)$, we first write G as a graph with $2|V(G)|$ directed edges, as in Fig. 9.3.

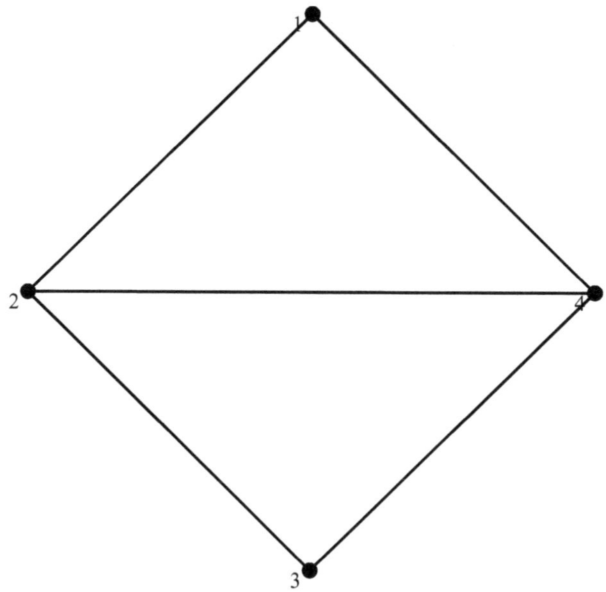

Fig. 9.2 Undirected simple example graph G with $N = 4$ vertices and $M = 5$ edges. Visual inspection shows that it has $\mathcal{C}_3(G) = 2$ (two 3-cycles), $\mathcal{C}_4(G) = 1$ (one 4-cycle), and $\mathcal{C}_k(G) = 0$ for all $k \neq 3,4$

From this new graph we then construct the edge graph $D(G)$, which for the present example has the topology shown in Fig. 9.4. Note that for our formulas (which are expressed solely in terms of quantities that are invariant under renaming all vertices, provided the adjacency matrix is transformed correspondingly), the labeling of the vertices in $V(D(G))$ is not important. We can now read of the entries of the adjacency matrix T of $D(G)$, yielding Fig. 9.5.

This explicit construction illustrates the subtle connection that exists between the cyclic structure of the original graph G and its associated edge graph $D(G)$, as expressed by the theorem of Kotani and Sunada.

The matrix T in (9.2) is diagonalized trivially (note that here we have $m = 10$):

$$\det(I - Tu) = u^{10} - 4u^7 - 2u^6 + 4u^4 + 4u^3 + u^2 - 4. \tag{9.21}$$

If for this present example (9.2) we then calculate, via identity (9.19), the coefficients of the Ihara zeta function, we obtain the following result:

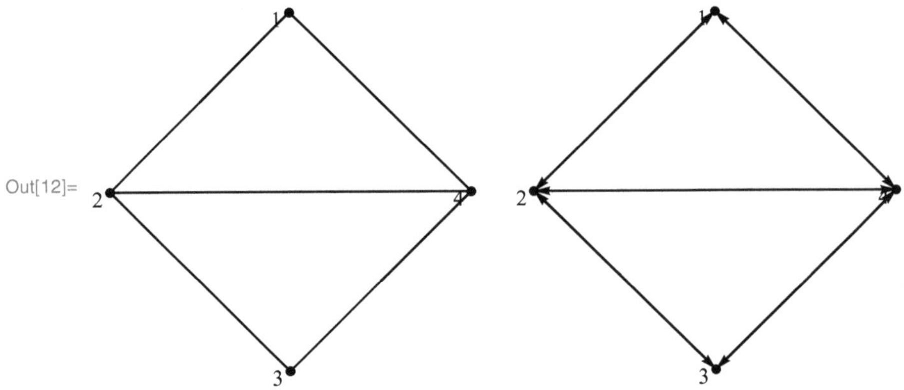

Fig. 9.3 First step in the construction of the edge graph associated with G: the interpretation of the undirected graph on the *left* as a graph with $2|E(G)|$ *directed* edges (on the *right*)

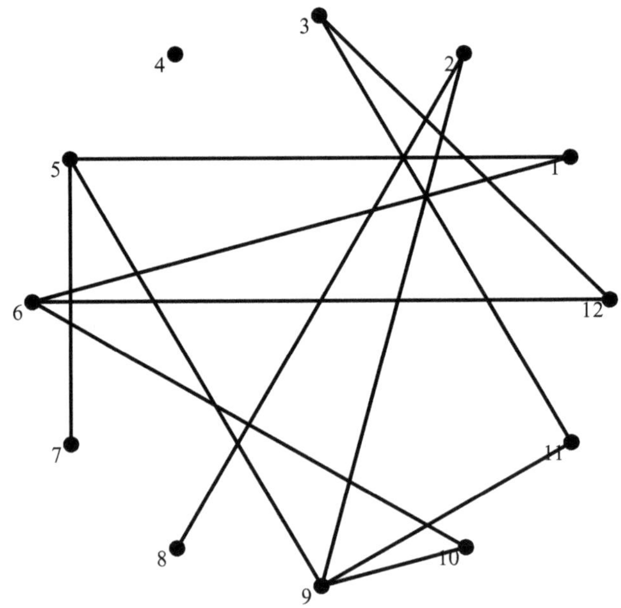

Fig. 9.4 The oriented line graph $D(G)$ corresponding to the undirected graph G of Fig. 9.2

$$
\begin{aligned}
c_1 &= 0,\\
c_2 &= \text{minus twice the number of primitive cycles of length 2 in } G = 0,\\
c_3 &= \text{minus twice the number of triangles in } G = -4,\\
c_4 &= \text{the number of primitive cycles of length 2}\\
&\quad + \text{twice the number of pairs of primitive cycles of length 2 that share an edge}\\
&\quad + \text{four times the number of edge-disjoint pairs off primitive cycles of length 2}\\
&\quad - \text{twice the number of squares in } G = -2.
\end{aligned}
$$

$$T = \begin{pmatrix} 0&0&0&0&1&1&0&0&0&0&0&0\\ 0&0&0&0&0&0&1&1&0&0&0&0\\ 0&0&0&0&0&0&0&0&0&0&1&1\\ 0&1&1&0&0&0&0&0&0&0&0&0\\ 0&0&0&0&0&0&1&0&1&0&0&0\\ 0&0&0&0&0&0&0&0&0&1&0&1\\ 1&0&1&0&0&0&0&0&0&0&0&0\\ 0&0&0&1&0&1&0&0&0&0&0&0\\ 0&0&0&0&0&0&0&0&0&1&1&0\\ 1&1&0&0&0&0&0&0&0&0&0&0\\ 0&0&0&1&1&0&0&0&0&0&0&0\\ 0&0&0&0&0&0&1&1&0&0&0&0 \end{pmatrix}$$

Fig. 9.5 The adjacency matrix T of the edge graph $D(G)$ associated with the graph G of Fig. 9.2

We conclude that there are two cycles of length 3 in G, and one cycle of length 4. With the edge zeta function, we have determined the basic cycle structure of the graph.

9.5 Conclusions

In this paper, we have surveyed the mathematical methodology available for determining the number of short walks and cycles in complex networks. In particular, we have discussed an elegant and computationally efficient method for quantifying the number of short cycles in a network, which is based on the graph zeta function. The application of these methods to real biological networks—protein interaction networks—will be the subject of a future study.

References

1. Albert, R., Barabási, A.L.: Statistical mechanics of complex networks. Rev. Mod. Phys. **74**, 47–97 (2002)
2. Dorogovtsev, S.N., Goltsev, A.V., Mendes, J.F.F.: Critical phenomena in complex networks. Rev. Mod. Phys. **80**, 1275–1335 (2008)
3. Newman, M.E.J.: Networks: An Introduction. Oxford University Press, Oxford (2010)
4. Ihara, Y.: Discrete subgroups of PL(2,Ks). Algebraic groups and discontinuous subgroups. In: Proceedings of Symposia in Pure Mathematics. Boulder, Colorado (1965)
5. Bass, H.: The Ihara–Selberg zeta function of a tree lattice. Int. J. Math. **3**(6), 717–797 (1992)
6. Stark, H.M., Terras, A.: Zeta functions of finite graphs and coverings. Adv. Math. **121**(1), 124–165 (1996)
7. Bollobás, B.: Random Graphs. Academic, New York (1985)
8. Watts, D.J., Strogatz, S.H.: Collective dynamics of small-world networks. Nature **393**, 440–442 (1998)

9. Bollobás, B., Riordan, O.M.: Handbook of Graphs and Networks: From the Genome to the Internet. Wiley-VCH, Weinheim (2003)
10. Coolen, A.C.C., Fraternali, F., Annibale, A., Fernandes, L., Kleinjung, J.: Modelling biological networks via tailored random graphs. In: Stumpf, M., Balding, D.J., Girolami, M. (eds.) Handbook of Statistical Systems Biology. Wiley, Chichester (2011)
11. Roberts, E.S., Schlitt, T., Coolen, A.C.C.: Tailored graph ensembles as proxies or null models for real networks II: results on directed graphs. J. Phys. A **44**, 275002 (2011)
12. Riemann, B.: Ueber die Anzahl der Primzahlen unter einer gegebenen Grösse, 1859. In: Gesammelte Werke. Teubner, Leipzig (1892) [Reprinted by Dover, New York (1953)]
13. Selberg, A.: On the estimation of Fourier coefficients of modular forms. In: Whiteman, A.L. (ed.) Theory of Numbers. Proceedings of the Symposium on Pure Mathematics, vol. 8, pp. 1–15. American Mathematical Society, Providence (1965)
14. Ihara, Y.: On discrete subgroups of the two by two projective linear group over p-adic fields. J. Math. Soc. Jpn. **18**, 219–235 (1966)
15. Hashimoto, K.: Zeta functions of finite graphs and representations of p-adic groups. In: Advanced Study in Pure Mathematics, vol. 15, pp. 211–280. Academic, New York (1989)
16. Kotani, M., Sunada, T.: Zeta functions of finite graphs. J. Math. Sci. Univ. Tokyo **7**, 7–25 (2000)
17. Scott, G., Storm, C.: The coefficients of the Ihara zeta function. Involve **1**(2), 979–991 (2008)
18. Grácio, C., Sousa Ramos, J.: Boundary maps and Fenchel-Nielsen-Maskit coordinates. Int. J. Bif. Chaos. **13**(7), 1949–1956 (2003)

Chapter 10
Independence Test of Chaotic Sequences

Zouhair Ben Jemaa, Daniele Fournier-Prunaret, and Safya Belghith

Abstract For the last several years, chaotic sequences have been considered and used in many applications to produce pseudorandom sequences; this has been motivated by the fact that chaotic systems are deterministic sources demonstrating some random features. Many research works have studied these features and analyzed and measured the randomness of these sequences. In particular, it has been shown that under some assumptions, a chaotic sequence can be a random source that delivers continuous random variables with invariant probability density. In many works it has been assumed that chaotic sequences generated by a nonlinear map and beginning with two different initial conditions are independent. The aim of this paper is to analyze the extent to which this is true and to measure this independence.

10.1 Introduction

Over the last several years, many searches have dealt with chaotic sequences; two aspects have been considered: theoretical and applications. In the theoretical aspect, it has been shown that a chaotic sequence, although it is generated by a deterministic process, could be considered the result of a random process [1–3]. Therefore, many statistical features of chaotic sequences such as auto- and cross-correlation [4, 5] and probability density function [6, 7] have been the subject of recent research. In the application aspect, chaotic sequences have been used in telecommunications, cryptography, and watermarking [8–14]; it has also been shown that this makes it possible to improve performance with respect to classical solutions. Specifically, in

Z. Ben Jemaa (✉) • S. Belghith
SysCom Laboratory, Ecole Nationale d'Ingénieurs de Tunis, Tunis, Tunisia
e-mail: zouhair.benjemaa@enit.rnu.tn; safya.belghith@enit.rnu.tn

D. Fournier-Prunaret
CNRS, LAAS, Université de Toulouse, Toulouse, France
e-mail: daniele.fournier@insa-toulouse.fr

C. Grácio et al. (eds.), *Nonlinear Maps and their Applications*, Springer Proceedings in Mathematics & Statistics 57, DOI 10.1007/978-1-4614-9161-3_10,

spread spectrum communication systems, chaotic sequences have been considered as multiple access spreading codes; in most of the cases in which the statistics of interferences between users were computed, it has been supposed that these sequences are independent [9, 10].

The aim of this paper is to analyze how much of this assumption is true and whether the answer depends on the initial conditions of the considered chaotic sequences and/or on the map generating them.

This paper is organized as follows; in Sect. 10.2, we will recall the notion of independence of two random variables and the Chi2 test of independence in both discrete and continuous cases; Sect. 10.3 deals with the random aspect of chaotic sequences; in Sect. 10.4, we will present the results of applying the Chi2 independence test to chaotic sequences for different initial conditions; in Sect. 10.5, we will analyze the influence of the considered chaotic maps on the Chi2 test, and then we will conclude and give some remarks in Sect. 10.6.

10.2 Independence of Random Time Series

Let X and Y be two random variables. X and Y are independent iff

$$P(X \in M, Y \in N) = P(X \in M)P(Y \in N) \tag{10.1}$$

for all Boreal sets M, N.

This is the mathematical definition of independence; but sometimes, we have the realizations of two random variables with no idea about their independence and we need to decide whether they are independent. This will be discussed in the next section.

10.2.1 Independence Test of Discrete Random Variables

Let $(x,y) = ((x_1,y_1),(x_2,y_2),\dots,(x_n,y_n))$ be n samples of two random variables X and Y taking values respectively in finite sets $A = \{a_1,a_2,\dots,a_k\}$ and $B = \{b_1,b_2,\dots,b_\ell\}$. We note that $p_i = P(X = a_i)$, $q_j = P(Y = b_j)$, and $r_{ij} = P(X = a_i, Y = b_j)$. As described above, X and Y are independent if and only if $p_i q_j = r_{ij}$ for all $1 \leq i \leq k$ and $1 \leq j \leq \ell$. Now we assume that no information about the independence of X and Y is available; and we want to test this independence using the sample (x,y). This is done by applying the Chi2 test [15, 16] described here. Let

$$\hat{p}_i = \frac{n_x^i}{n}, \hat{q}_j = \frac{n_y^j}{n}, \hat{r}_{ij} = \frac{n_{xy}^{ij}}{n}, \tag{10.2}$$

where n_x^i and n_y^j are the numbers of elements in the two sets of integers $\{t, x_t = a_i\}$ and $\{t, y_t = b_j\}$, n_{xy}^{ij} is the number of elements in the set $\{(t,s), x_t = a_i, y_s = a_j\}$ of pairs of integers. The χ^2 test statistic with degree of freedom $DF = (k-1)(\ell-1)$ of the two sequences x and y is defined by

$$\chi^2 = n \sum_{i=1}^{k} \sum_{j=1}^{\ell} \frac{(\hat{r}_{ij} - \hat{p}_i \hat{q}_j)^2}{\hat{p}_i \hat{q}_j}. \tag{10.3}$$

For a given pair of sequences (x,y), χ^2 is used to decide as to their independence if (x,y) are random χ^2 is a continuous random variable with a probability density $\rho_{DF}(x)$ depending on the degree of freedom DF. In Fig. 10.1 are plotted curves of $\rho_{DF}(x)$ for different values of DF.

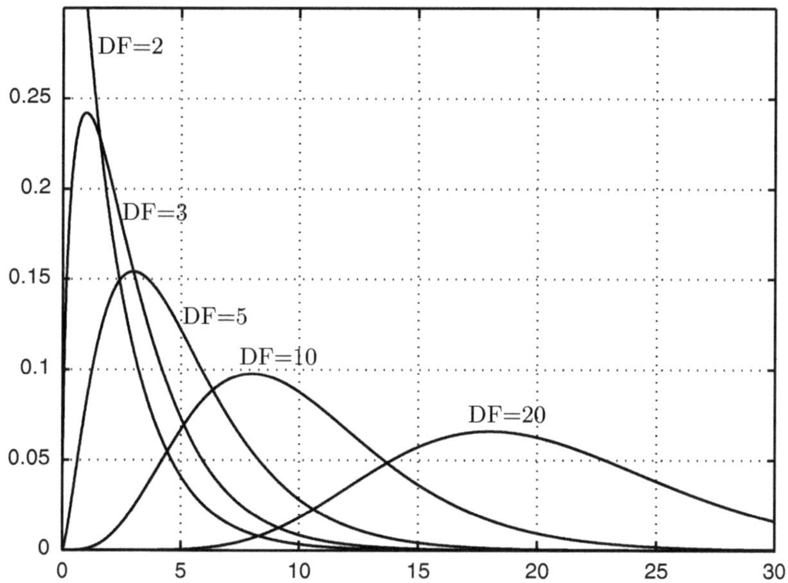

Fig. 10.1 Probability density χ^2 for different degree of freedom DF

For a given sample (x,y), the computed χ^2 test statistic value of (10.3) is a nonzero positive number; the question is how we can decide whether x and y are independent. The answer to this question depends on our tolerance in taking this decision. Thus for a given DF, we choose a significance level $\varepsilon > 0$; a threshold α is defined by

$$p(\chi^2 > \alpha) = \varepsilon. \tag{10.4}$$

This is illustrated in Fig. 10.2, where we have plotted some probability density of χ^2; we decide that x and y are independent if $\chi^2 < \alpha$. The area ε is the probability of having a negative test of independence, whence if the threshold α increases, the tolerance is greater and the probability ε is lower.

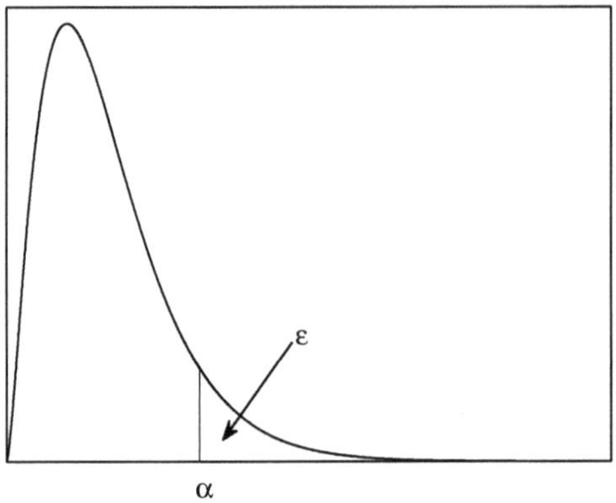

Fig. 10.2 Threshold and tolerance of Chi2 test

10.2.2 Independence Test of Continuous Random Variables

Let X and Y be two continuous random variables taking values in two intervals $[a,b]$ and $[c,d]$. Let $(x,y) = ((x_1,y_1),(x_2,y_2),\dots,(x_n,y_n))$ be n samples of the two random variables X and Y. To test the independence of x and y, the intervals $[a,b]$ and $[c,d]$ are partitioned into two sets $P_{ab} = \left\{\Delta_i = [a + \frac{i-1}{k}(b-a), a + \frac{i}{k}(b-a)]\right\}$ and $P_{cd} = \left\{\Lambda_j = [c + \frac{j-1}{\ell}(d-c), c + \frac{j}{\ell}(d-c)]\right\}$ for all $1 \le i \le k$ and $1 \le j \le \ell$. The χ^2 test statistic of the sample (x,y) is defined as above with the difference that n_x^i and n_y^j are the numbers of elements in the two sets $\{t, x_t \in \Delta_i\}$ and $\{t, y_t \in \Lambda_j\}$ respectively, n_{xy}^{ij} is the number of elements in the set $\{(t,s), x_t \in \Delta_i, y_s \in \Lambda_j\}$. We consider the example of two uniform random variables in the intervals $[a,b] = [c,d] = [0,1]$; to test the independence of these two variables, we consider $n = 10{,}000$ samples (x,y) and $k = \ell = 6$. We obtain the chi2 histogram represented in Fig. 10.3. It is worth noting that this histogram is plotted by considering the $n = 10{,}000$ computed values of the χ^2 statistic and a step equal to 2; it was normalized such that we obtain the same scale as the probability density $\rho_{25}(x)$. From the histogram, we can deduce that the decision on the independence strongly depends on the realization (x,y).

10.3 Randomness of Chaotic Sequences

In this section, we consider the stochastic aspect of sequences generated by a chaotic map. We will see how we can generate a probability density using a deterministic chaotic transformation. Let us begin by recalling some characteristics of the probability density of a chaotic function. Let

$$f : [0,1] \to [0,1] \tag{10.5}$$

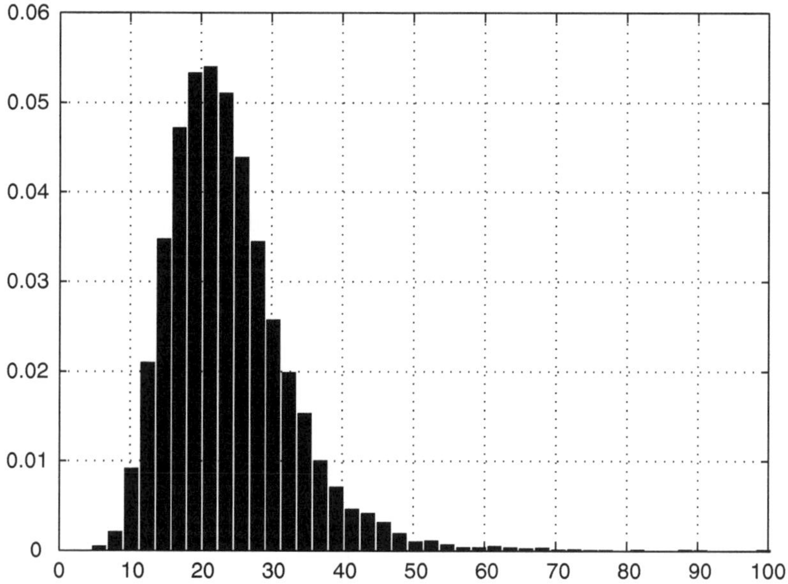

Fig. 10.3 Histogram of χ^2 in the case of the independence test of two uniform random variables

be a one-dimensional chaotic map. A trajectory x_n, $n = 0, 1, 2, \ldots$, is defined by an initial condition x_0 and the recurrence $x_{n+1} = f(x_n)$. We suppose that the initial condition x_0 is a continuous random variable with probability density noted by ρ_0. We denote by ρ_k the probability density of x_k obtained after k iterations of the map f. The probability density ρ_k converges to a probability density ρ_f that is invariant by iterating f [17]; i.e., if x is a random variable with probability density ρ, it is the same for $f(x)$. To illustrate this idea, we consider a set of N initial conditions $I_0 = \{x_0^1, x_0^2, \ldots, x_0^N\}$. To every initial condition x_0^i there corresponds a trajectory $x_0^i, x_1^i, \ldots, x_n^i, \ldots$; by applying the map f to the elements of I_0, we obtain a set $I_1 = \{x_1^1, x_1^2, \ldots, x_1^N\}$, and if we iterate successively the map f, we obtain a sequence of sets $I_k = \{x_k^1, x_k^2, \ldots, x_k^N\}$. To illustrate this idea, we considered the logistic map

$$L : [0,1] \to [0,1]\,;\; L(x) = 4x(1-x). \tag{10.6}$$

In Fig. 10.4, we plotted the histograms h_0, h_2, h_5, and h_{20} of the sets I_0, I_5, I_10, and I_∞. We can see that ρ_k converges to the invariant probability density $h_\infty(x) = \frac{1}{\pi\sqrt{x(1-x)}}$ [17].

10.4 Independence Test of Two Chaotic Sequences

Let x_0 and y_0 be two random initial conditions and $x_0, x_1, \ldots, x_n$ and let $y_0, y_1, \ldots, y_n$ be the two sequences beginning with x_0 and y_0. In Fig. 10.5, we have represented N realizations of each sequence x and y.

To test the independence of two sequences x and y, we proceed following two approaches; for the first, we consider two initial conditions chosen randomly and test whether the two sequences beginning with these two initial conditions satisfy the independence test. In Fig. 10.5, an example pair of sequences are put in elliptical

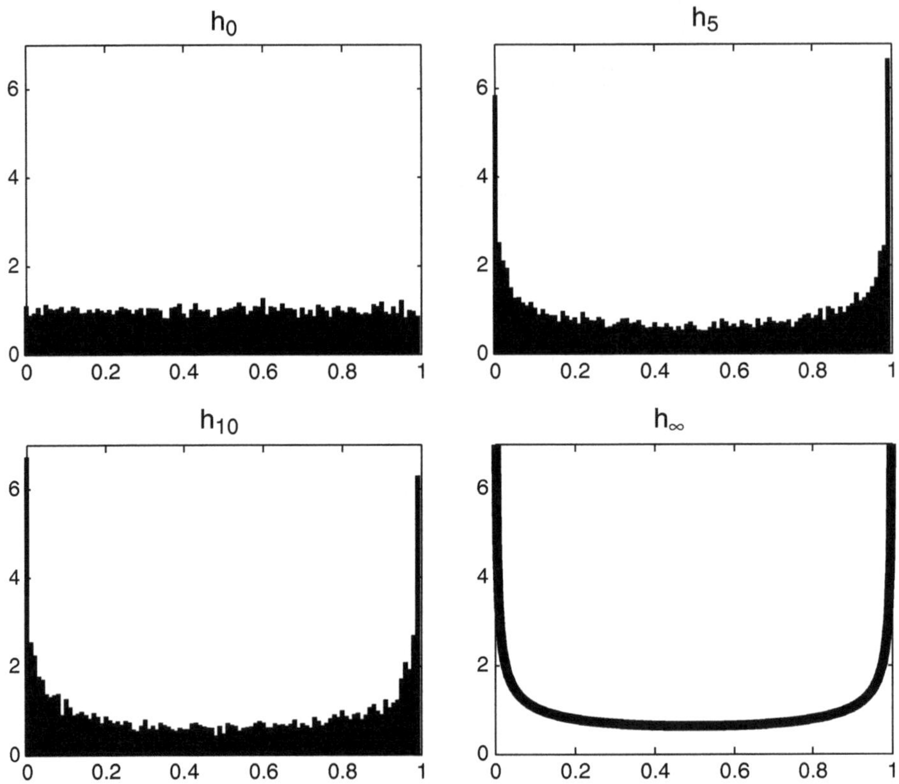

Fig. 10.4 Probability density evolution under application of the map (10.6)

boxes; for the second approach, we consider two sets of initial conditions and analyze the evolution of independence of these two sets by iterating them successively using the transformation (10.6), in Fig. 10.5. An example pair of sequences are put in rectangular boxes.

- **Approach 1**
  Let $x = x_0, x_1, \ldots, x_n$ and $y = y_0, y_1, \ldots, y_n$ be two sequences generated by iterating a chaotic map on the interval $[0,1]$. As done for a random variable, we aim in

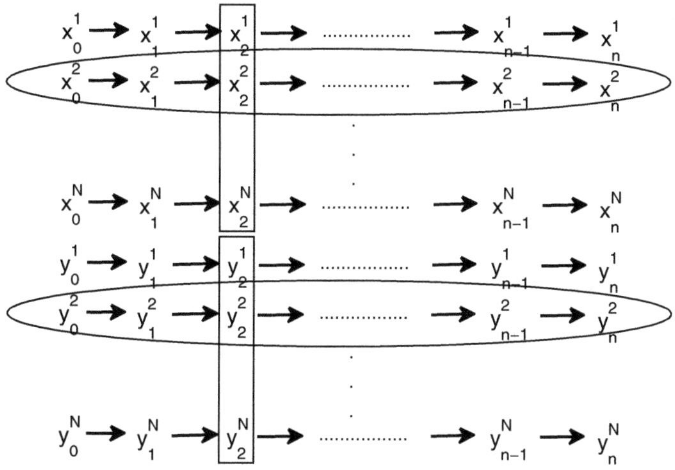

Fig. 10.5 Two approaches of chaotic sequences Chi2 test

this section to answer the following questions: are x and y independent? does the answer to this question depend on the choice of (x_0, y_0)?
We have the following classical result: if x_0 and y_0 are independent, then $x_k = f^k(x_0)$ and $y_k = f^k(y_0)$ are also independent, and no independence test is needed. Thus we suppose that x_0 and y_0 are random but not independent variables; precisely, we consider x_0 a random variable and $y_0 = x_0 + \delta x_0$; this allows us to analyze the sensitivity to initial conditions in its stochastic aspect. We consider here the map (10.6) and plot in Fig. 10.6 the χ^2 test statistic of the pair (x, y), where $x = x_0, \dots, x_n$ and $y = y_0, \dots, y_n$ for different realizations of x_0 and $y_0 = x_0 + \delta x_0$ and for $\delta x_0 = 0.01$. The first thing that we can deduce is that χ^2 strongly depends on the initial conditions (x_0, y_0); moreover, depending on the threshold α and x_0, we can decide whether the two sequences are independent.

- **Approach 2**
  Let $x_0^1, x_0^2, \dots, x_0^N$ be a set of samples of random initial condition x_0 and let $y_0^1, y_0^2, \dots, y_0^N$ be such that $y_0^i = x_0^i + \delta x_0$, $\delta x_0 = 0.01$. This approach consists in testing the independence of the sets $x_k^1, x_k^2, \dots, x_k^N$ and $y_k^1, y_k^2, \dots, y_k^N$; $x_k^i = f^k(x_0^i)$. The result is shown in Fig. 10.7. In this case also, the decision strongly depends on the iteration number, and for a given threshold, from one iteration to the next we can pass from independent to dependent sequences.

10.5 Statistics of χ^2 of Chaotic Sequences Generated by Different Chaotic Maps

We saw in the previous section that the decision on independence strongly depends on the initial conditions of the two sequences; in this section, we consider the following problem, with the same hypotheses of the previous section, and if we

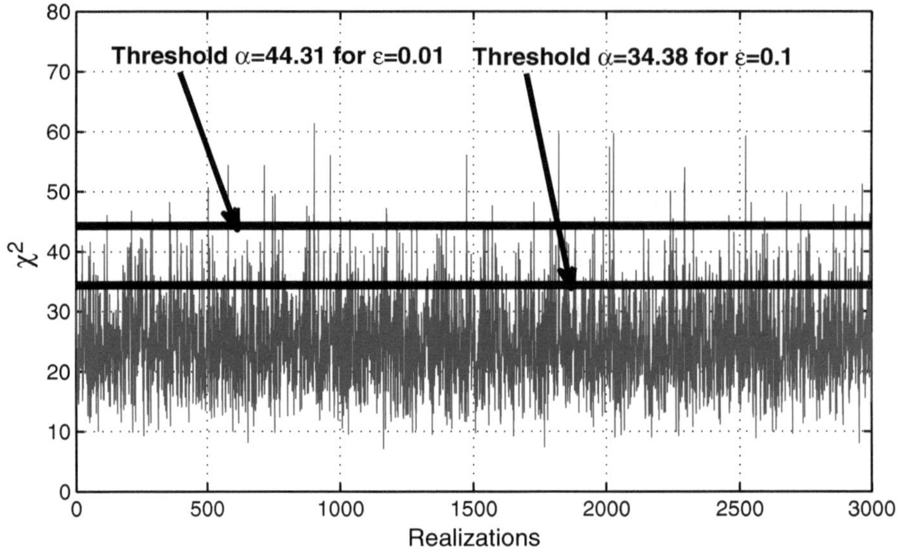

Fig. 10.6 χ^2 of $x_0,\ldots,x_n$ and $y_0,\ldots,y_n$ for different realizations of x_0 and $y_0 = x_0 + \delta x_0$ for the logistic map case ($\delta x_0 = 0.01$, $n = 10{,}000$, and $k = l = 6$)

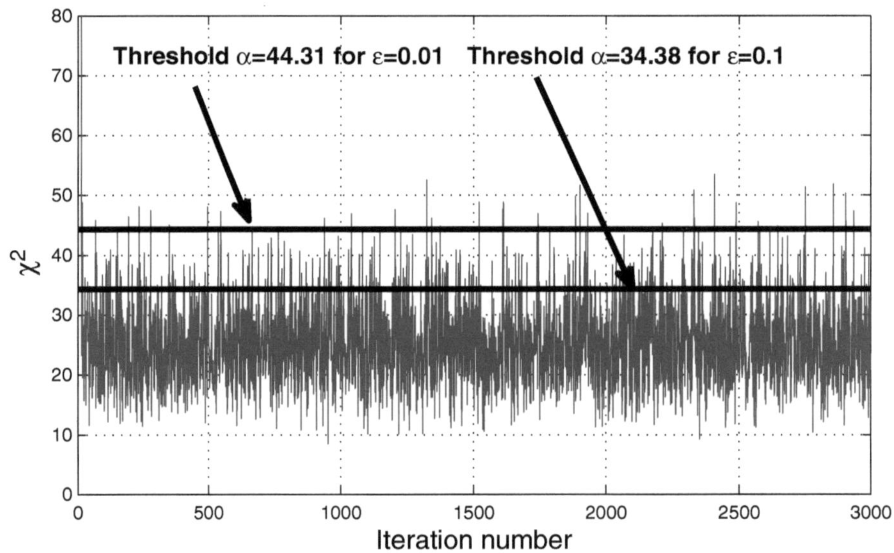

Fig. 10.7 (x,y) independence evolution for the logistic map case ($\delta x_0 = 0.01$, $N = 1{,}000$ pairs of initial conditions, and $k = l = 6$)

choose randomly the pair of initial conditions $(x_0, x_0 + \delta x_0)$, does the probability of obtaining a positive test of independence depend on the considered chaotic map? To answer to this question, we considered the following maps:

- The random case, in which x and y are random vectors uniformly distributed on the interval $[0,1]$. This case will be used as a reference.
- The logistic map (10.6).
- The skew-tent map with various parameters λ:

$$g_1(x) = \begin{cases} \frac{2}{\lambda-1}x - \frac{1+\lambda}{\lambda-1} & \text{if } \lambda < x \leq 1, \\ \frac{2}{\lambda+1}x - \frac{\lambda-1}{\lambda+1} & \text{if } -1 \leq x \leq \lambda. \end{cases} \tag{10.7}$$

- The map

$$S(x) = 2.6\sin(x+2.1)^2. \tag{10.8}$$

- The Bernoulli map

$$g_2(x) = \text{mod}\,(3x, 1). \tag{10.9}$$

These maps are represented in Fig. 10.8. As done above, we consider the following two approaches:

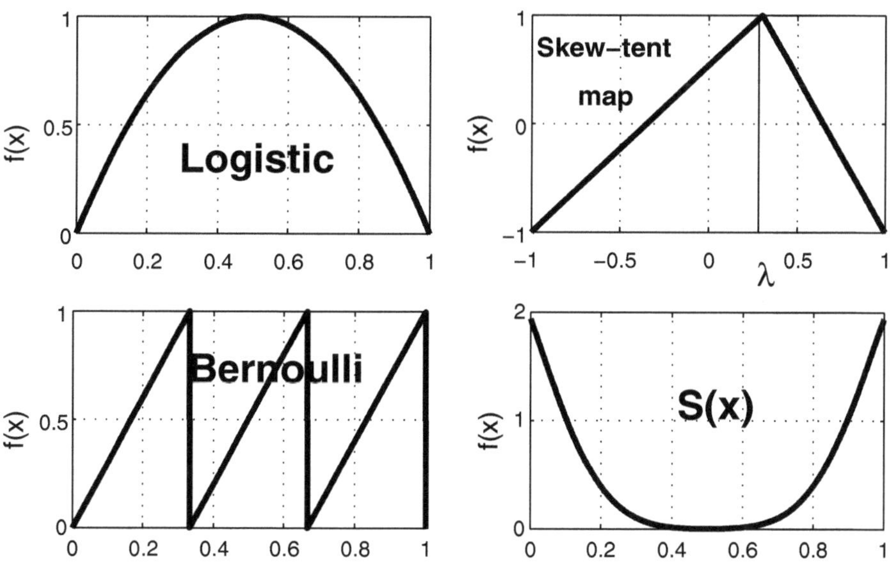

Fig. 10.8 Different considered maps

- **Approach 1**
  To try to answer this question according to approach 1, we compute the mean of the χ^2 test statistic and plot it versus the number of realizations. Moreover, we consider the case in which x and y are random vectors uniformly distributed on the interval $[0,1]$. It is worth noting that in the independence case, the theoretical

values of the χ^2 average and variance are DF and $2DF$, respectively, where DF is the degree of freedom; in our case, $DF = 25$, and the expected values are 25 and 50. In Fig. 10.9, we can see that the logistic map (10.6) and the skew-tent (10.7) for the parameter $\lambda = 0.1$ allow an average χ^2 close to the theoretical average, and better than those of the other maps (10.7) for the parameter $\lambda = 0.5$, with $S(x)$ given by (10.8) and Bernoulli (10.9). For the variance, the Bernoulli is the best; and the maps (10.7) for the parameter $\lambda = 0.5$ and $S(x)$ have the highest variance; this is illustrated in Fig. 10.10. Results are summarized in Table 10.1.

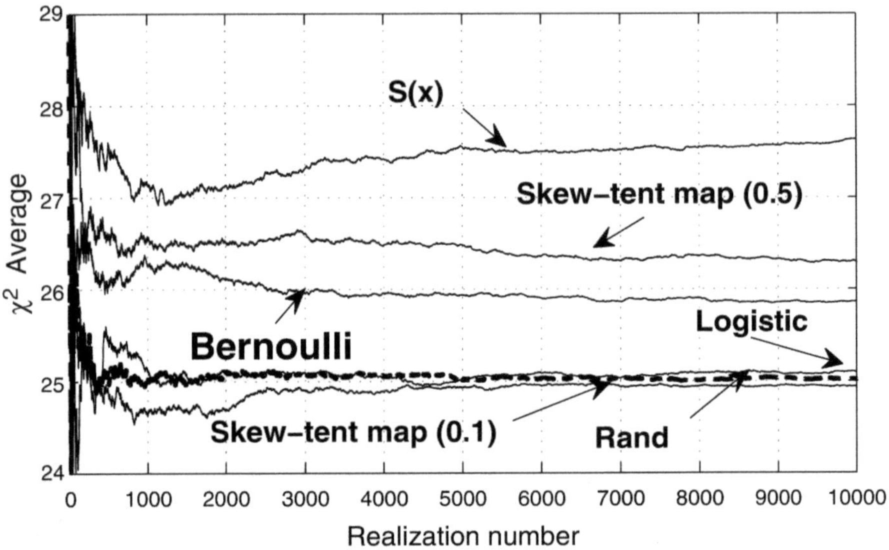

Fig. 10.9 χ^2 average versus realization number ($\delta x_0 = 0.01$, $n = 10{,}000$, and $k = l = 6$)

These results can be explained by looking at Fig. 10.11, where we plotted two chaotic sequences generated by the chaotic map (10.8). We zoomed at a point E_1 where the two trajectories fall close to each other. Here we can see that the two sequences remain correlated for a finite time before they again appear uncorrelated (point E_2). This situation does not exist in the random case. This explanation could also be applied to compare the results produced by the different chaotic sequences. Indeed, the stochastic and topological properties of the considered chaotic maps determine the frequency at which two chaotic sequences come close to each other, and when this happens, during how many iterations they remain correlated.

- **Approach 2**
  In this case, we can deduce from Figs. 10.12 and 10.13 that the average and the variance are almost the same for all the considered cases. Results are summarized in Table 10.2. The case of the Bernoulli map (10.9) seems surprising. The high values found for average and variance of χ^2 are due to the specificity of this

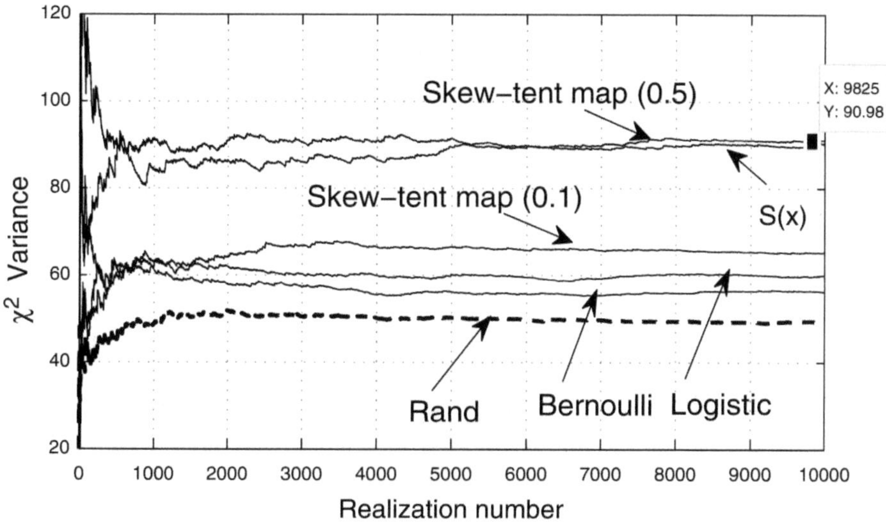

Fig. 10.10 χ^2 variance versus realization number ($\delta x_0 = 0.01$, $n = 10{,}000$, and $k = l = 6$)

Table 10.1 Statistics of χ^2 for different chaotic maps: approach 2

Map	Average	Variance
Independence	25	50
Random	25.02	49.44
Skew-tent map (0.1)	24.9	65.26
Logistic	25.1	60.02
Bernoulli	25.86	56.3
skew-tent map (0.5)	26.3	91
S(x)	27.6	90.3

map. Indeed, we can easily show that for x_0, we have $y_0 = x_0 + \delta x_0$, $x_l = f^l(x_0)$, and $y_l = f^l(y_0)$; $y_l - x_l$ takes only the two values $3^l\delta x_0 - E(3^l\delta x_0)$ and $3^l\delta x_0 - E(3^l\delta x_0) - 1$, where $E(x)$ is the floor of x. Thus the distance between the two sequences takes only two values that do not depend on the pair (x_0, y_0) but only on the initial perturbation δx_0; this gives a high value of χ^2.

10.6 Conclusion

In this paper, we have considered the problem of independence of two chaotic sequences and essentially showed that we cannot conclude the existence of such independence absolutely. Indeed, by considering the Chi2 test, we have shown that this independence strongly depends on the two dependent initial conditions of the two considered sequences. We also showed that this independence depends on the

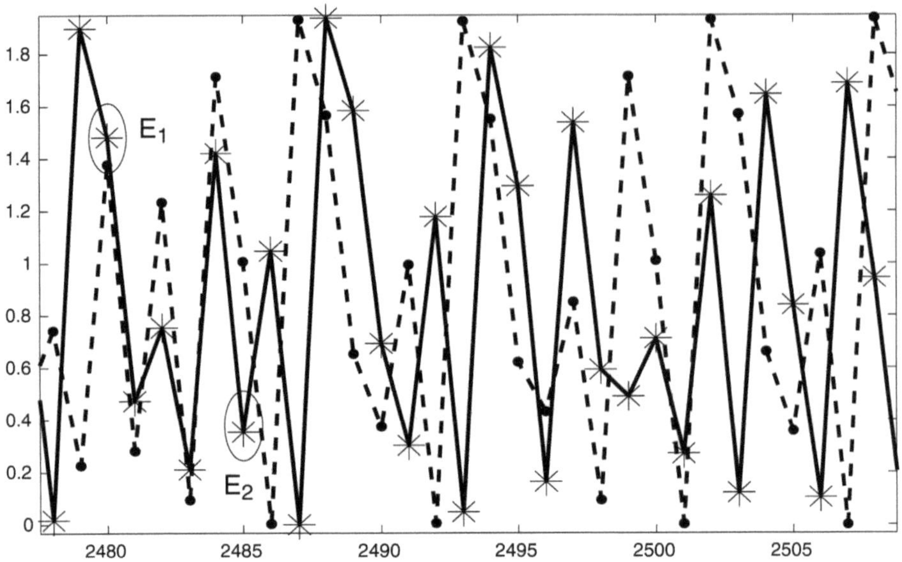

Fig. 10.11 Evolution of two chaotic sequences

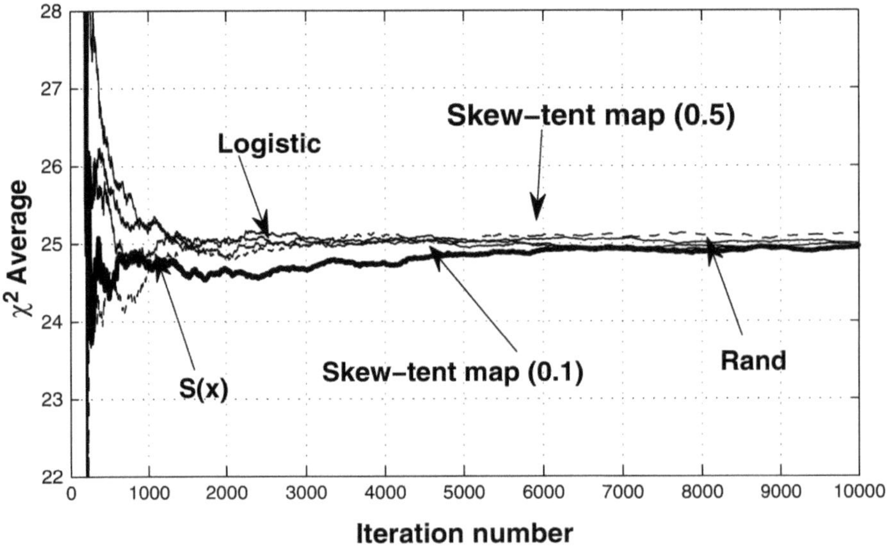

Fig. 10.12 χ^2 average versus iteration number ($\delta x_0 = 0.01$, $N = 1000$ pairs of initial conditions, and $k = l = 6$)

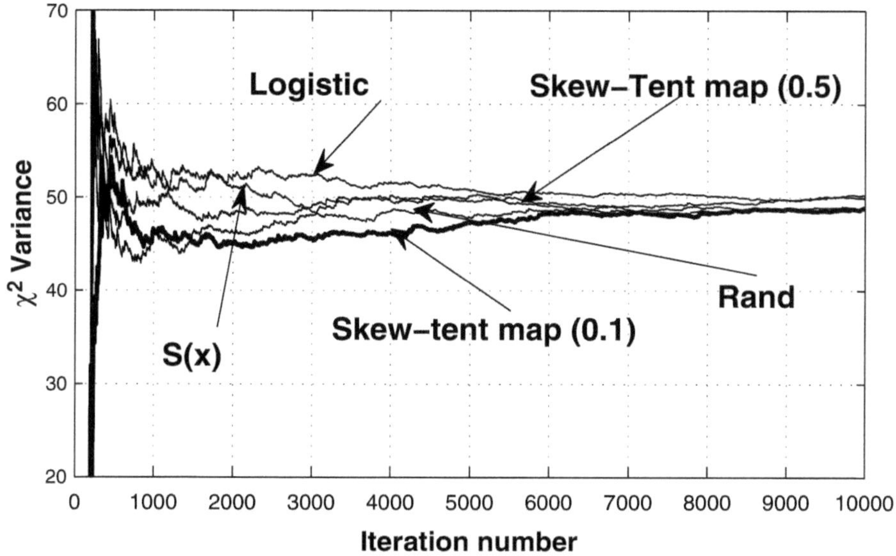

Fig. 10.13 χ^2 average versus iteration number ($\delta x_0 = 0.01$, $N = 1000$ pairs of initial conditions, and $k = l = 6$)

Table 10.2 Statistics of χ^2 for different chaotic maps: approach 2

Map	Average	Variance
Independence	25	50
Random	24.99	50.23
skew-tent map (0.1)	24.96	48.91
Logistic	24.96	49.89
Bernoulli	2,000	364,300
skew-tent map (0.5)	25.12	48.68
S(x)	24.92	48.72

length of the two sequences. To analyze the influence of the considered chaotic map, we computed the average and the variance of χ^2 over initial conditions and showed that these statistics depend on the chaotic map. It would be interesting to deepen this study by looking for a relationship between the statistical and topological properties of the considered chaotic maps and the results that have been obtained here with them.

Acknowledgements We thank Dr. Viktor Avrutin for interesting and helpful discussions during the NOMA workshop.

References

1. Pareschi, F.: Chaos based random number generators: Monolithic implementation, testing and applications. Ph.D. thesis, Università di Bologna, December (2006)
2. Fournier-Prunaret, D., Taha, A.: Chaotic sequences as pseudo-random sequences obtained from 2-dimensional maps. IJAMAS **26**(2), 92–105 (2012)
3. Zheng, F., Tian, X., Song, L., Li, X.: Pseudo-random sequences generator based on the generalised Hénon map. J. China Univ. Posts Telecommun. **15**(3), 64–68 (2008)
4. Kohda, T., Fujisaki, H., Ideue, S.: On distributions of correlation values of spreading sequences based on Markov information sources. In: Proceedings of ISCAS'2000, pp. 225–228 (2000)
5. Kohda, T., Tsuneda, A.: Explicit evaluations of correlation functions of Chebyshev binary and bit sequences based on Perron–Frobenius operator. IEICE Trans. Fundam. Electron. Commun. Comput. Sci. **E77-A**, 1794–1800 (1994)
6. Götz, M., Schwarz, W.: Spectral decomposition of the Frobenius–Perron operator and higher-order correlation functions of one-dimensional chaotic systems with fully stretching maps. In: Proceedings of European Conference on Circuit Theory Design, pp. 1253–1258 (1997)
7. Kohda, T., Fujisaki, H.: The Jacobian elliptic Chebyshev rational maps can generate sequences of iid random variables. In: Proceedings of ISCAS'2000, pp. 221–224 (2000)
8. Chargé, P., Fournier-Prunaret, D., Guglielmi, V.: Features analysis of a parametric PWL chaotic map and its utilization for secure transmissions. Chaos Soliton. Fract. **38**(5), 1411–1422 (2008)
9. Mazzini, G., Setti, G., Rovatti, R.: Chaotic complex spreading sequences for asynchronous DS-CDMA-Part I: System modeling and results. IEEE Trans. Circuit Syst. I **44**, 937–947 (1997)
10. Rovatti, R., Setti, G., Mazziniv, G.: Chaotic complex spreading sequences for asynchronous DS-CDMA-Part II: Some theoretical performance bounds. IEEE Trans. Circuit Syst. I **45**, 496–506 (1998)
11. Broomhead, D.S., Huke, J.P., Muldoon, M.R.: Codes for spread spectrum, applications generated using chaotic dynamical systems. Dyn. Stab. Syst. **14**(1), 95–105 (1999)
12. Kaddoum, G.: Contributions á l'amélioration des systèmes de communication multi-utilisateurs par chaos: synchronisation et analyse des performances. Ph.D. thesis, INSA Toulouse (2008)
13. Caroll, T.L.: Spread spectrum sequences for unstable periodic orbits. IEEE. Trans. Circuit Syst. Part I **47**(4), 443–447 (2000)
14. Ben Jemaa, Z., Marcos, S., Belghith, S.: Performances of MMSE receiver when using super stable orbits based spreading sequences in DS-CDMA System. In: Proceedings of Eusipco (2000)
15. Hartmann, A.K.: Introduction to Randomness and Statistics. Wiley, New York (2000)
16. Chase, W., Brown, F.: General Statistics. World Scientific, Singapore (2009)
17. Cencini, M., Cecconi, F., Vilpiani, A.: Chaos, from Simple Models to Complex Systems. World Scientific Publishing Co. Pte. Ltd, Singapore (2010)

Chapter 11
Stability Transformation Method for Unstable Periodic Orbits and Its Realization

Yuki Kawai and Tadashi Tsubone

Abstract In this paper, we consider a stability transformation method (STM) of unstable periodic orbits (UPOs) of a simple chaotic circuit with nonlinear characteristics. In previous work, we have proposed a controlling method that transforms the stability of any unstable fixed point (UFP) embedded in the chaos attractor without the information of the location and have demonstrated a procedure for obtaining the transformed systems. However, there is a problem of robustness when this system applies to UPOs. We propose an extended method for the periodic orbits and illustrate the theoretical analysis on an example of the Bernoulli shift map. Also, we confirm some results by laboratory measurements.

11.1 Introduction

Analyzing stability of nonlinear systems is one of the most important and basic aspects of nonlinear problems and engineering applications. A nonlinear system often exhibits an unstable oscillatory phenomenon called chaos. Control of nonlinear systems with chaos has been actively researched, and several methods to stabilize unstable periodic orbits (UPOs) that exist in a chaotic attractor have been proposed. Generally, the stabilization of UPOs is called *control of chaos* [1]. Delayed feedback control [2] (DFC) is a representative example.

Delayed feedback control can stabilize unknown steady states. This property means that DFC can discover unknown UPOs. The problem of stabilization of unknown steady states has received some attention, and DFC has been used for approaches to the solution of this issue [3–5]. However, DFC has a limitation, called the odd number limitation [6, 7]. Much research has been undertaken to overcome this limitation (for example, [8, 9]), but simpler approaches would be convenient.

Y. Kawai • T. Tsubone (✉)
Nagaoka University of Technology, 1603-1, Kamitomioka, Nagaoka, Niigata 940-2188, Japan
e-mail: ykawai@stn.nagaokaut.ac.jp; tsubone@vos.nagaokaut.ac.jp

C. Grácio et al. (eds.), *Nonlinear Maps and their Applications*, Springer Proceedings in Mathematics & Statistics 57, DOI 10.1007/978-1-4614-9161-3_11,

Other approaches to the stabilization of unknown steady states can be found in the literature on prediction-based feedback control, adaptive dynamic controllers, state observer approaches, washout filter-aided feedback control, and so on [10–14]. In these works, some theoretical and numerical results have been provided, but more detailed discussions for implementations to real systems are needed.

On the other hand, the stability transformation method [15, 16] (STM) was proposed in order to detect the location of unknown UPOs of dynamical systems that exhibit chaotic behavior. This method seeks a transformed dynamical system with a stable periodic orbit that has exactly the same location as a UPO of the target chaotic system. STM has no limitation such as DFC. In previous work, we applied a controlling method based on STM to a piecewise-constant chaos generator and provided theoretical analysis of the stability and the domain of attraction [17, 18]. However, the basic mechanism of the STM to detect l-period UPOs boils down to finding fixed points of the l-fold composition of the return map. Hence, there are problems around robustness and its realization because the composite map has generally a huge enlargement factor in the case of high-periodic orbits. It is important to consider the robustness mechanism and to tackle this problem, because some noise in a dynamical system may affect the characteristics of systems [19]. Therefore, we propose an extension of the controlling method based on STM to a more robust method of detecting periodic orbits, and we analyze the stability of the transformed system.

In Sect. 11.2, we introduce our previous stability transformation system [17] for fixed points. In Sect. 11.3, we present a proposed stability transformation system for periodic orbits. In Sect. 11.4, we introduce a piecewise-constant system as the basic chaotic system, and some theoretical results are verified in an experimental circuit. Finally, conclusions and future work are given in Sect. 11.5.

11.2 Previous Controlling Method Based on STM

In this section, we present our previous method [17]. The transformed system can be described as follows:

$$\begin{aligned}
&\begin{cases} x_0(n+1) = F(x_0(n) - K(x_1(n) - x_0(n))) \\ x_1(n+1) = x_0(n) - K(x_1(n) - x_0(n)) \end{cases} && \text{for } n = kl, \\
&\begin{cases} x_0(n+1) = F(x_0(n)) \\ x_1(n+1) = x_1(n) \end{cases} && \text{otherwise,}
\end{aligned} \tag{11.1}$$

where F is a nonlinear function, $x_0(n)$ and $x_1(n)$ are state variables at time n, K is a control gain, and k is any positive integer. Here, the return map from $(x_0(kl), x_1(kl))$ to $(x_0((k+1)l), x_1((k+1)l))$ can be given by

$$\begin{aligned}
x_0((k+1)l) &= F^l(x_0(kl) - K(x_1(kl) - x_0(kl))), \\
x_1((k+1)l) &= x_0(kl) - K\{x_1(kl) - x_0(kl)).
\end{aligned} \tag{11.2}$$

In order to generate stabilized UPOs and analyze the stability of the system around periodic points, let us derive a 1-D return map of the state variable $x_1(n)$. Observing the system (11.2) yields $x_0(n) = F^l(x_1(n))$. Hence the following 1-D return map can be given:

$$x_1((k+1)l) = F^l(x_1(kl)) - K(x_1(kl) - F^l(x_1(kl))),\\ (x_0(n) = F^l(x_1(n))). \tag{11.3}$$

A point p_i is said to be an l-periodic point if $p_i = F^l(p_i)$, $i = 1,2,3,\cdots,l$, and $p_i \neq F^k(p_i)$, $1 < k < l-1$. The stability condition can be described as follows as is the case in $l = 1$

$$\left| \left. \frac{\partial x_1((k+1)l)}{\partial x_1(kl)} \right|_{x_1(kl)=p_i} \right| < 1. \tag{11.4}$$

Let $A = \left. \frac{\partial F^l(x_1(n))}{\partial x_1(n)} \right|_{x_1(n)=p_i}$. The gain of a controller for a superstable condition is given as follows:

$$K = A(1-A)^{-1}. \tag{11.5}$$

Figure 11.1a shows the transformed system with stabilized unstable fixed points (UFPs). Here, the dashed line represents the original chaos map F, and the bold line is the shape of (11.3) in the superstable condition. In this example, the chaos map is a well-known Bernoulli shift map. Similarly, Fig. 11.1b shows the transformed system with stable 2-periodic UPO. In this figure, the dashed line represents the original composite map F^2. However, the composite map has generally a huge enlargement factor in the case of a high-periodic orbit.

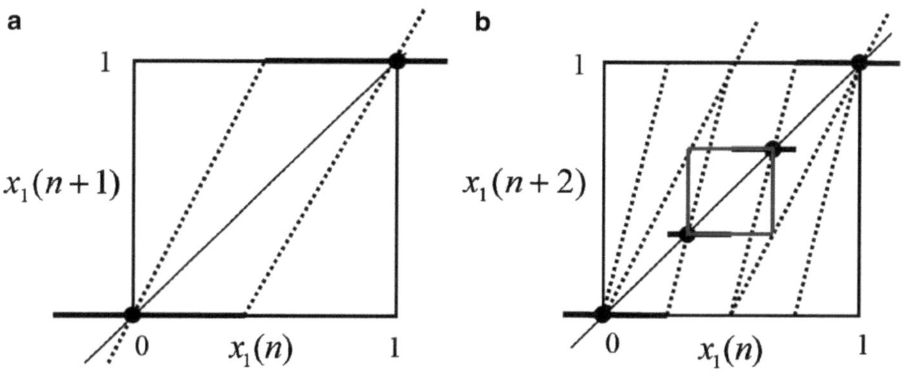

Fig. 11.1 Example of base chaos system and transformed system, (**a**) $(l = 1, K = 2)$, (**b**) $(l = 2, K = \frac{4}{3})$

11.3 Proposed Method Based on STM for Periodic Orbits

We have proposed an extension of the previous method for detecting periodic orbits. The proposed transformed system can be described as follows:

$$\begin{aligned}
x_0(n+1) &= F(x_0(n) - K_0(x_l(n) - x_0(n))) \\
x_1(n+1) &= x_0(n) - K_0(x_l(n) - x_0(n)) \\
x_i(n+1) &= x_{i-1}(n) - K_0 \prod_{j=0}^{i-2} K(n-j)(x_l(n) - x_0(n)) \\
(i &= 2, 3, \cdots, l),
\end{aligned} \tag{11.6}$$

where l is a period of the target UPO, K_0 is a constant control gain, and $K(n)$ is a time-variant control gain. This system is an $(l+1)$-dimensional system depending on the period of a targeted l-periodic UPO. Note that in the case of $l = 1$, this means that a targeted UPO is to be a fixed point. Then the system (11.6) can be identified with the previous system (11.1).

In order to analyze the stability of the system, we derive the following local linearized systems around the periodic points:

$$\begin{pmatrix} x_0(n+1) \\ x_1(n+1) \\ \vdots \\ x_l(n+1) \end{pmatrix} = A_i \begin{pmatrix} x_0(n) \\ x_1(n) \\ \vdots \\ x_l(n) \end{pmatrix}; \tag{11.7}$$

here A_i are the Jacobi matrices around the periodic points, described by

$$A_i = \begin{pmatrix}
a_i(1+K_0) & 0 & 0 & \cdots & 0 & -a_i K_0 \\
1+K_0 & 0 & 0 & \cdots & 0 & -K_0 \\
K_0 K(n) & 1 & 0 & \cdots & 0 & -K_0 K(n) \\
K_0 K(n) K(n-1) & 0 & 1 & \cdots & 0 & -K_0 K(n) K(n-1) \\
\vdots & \vdots & \vdots & \ddots & \vdots & \vdots \\
K_0 \prod_{j=0}^{l-1} K(n-j) & 0 & 0 & \cdots & 1 & -K_0 \prod_{j=0}^{l-1} K(n-j)
\end{pmatrix}, \tag{11.8}$$

where a_i is the slope of the original chaos map on the periodic points p_i:

$$a_i = \left. \frac{\partial F(x(n))}{\partial x(n)} \right|_{x(n)=p_i}. \tag{11.9}$$

Note that the determinant of A_i must be equal to zero regardless of the value of the parameters:

$$\det A_i = (-1)^{l-1} a_i \begin{vmatrix} 1+K_0 & -K_0 \\ 1+K_0 & -K_0 \end{vmatrix} = 0. \tag{11.10}$$

Also, we determine the time-variant control gain $K(n)$ as follows:

$$K(n) = \begin{cases} K_1 \text{ for } n = lk+1 \\ K_2 \text{ for } n = lk+2 \\ \quad \vdots \\ K_l \text{ for } n = lk \end{cases} \tag{11.11}$$
$$(k = 0, 1, 2, \dots).$$

In order to evaluate the stability of the periodic orbit, we have to analyze the matrix product $M = A_l A_{l-1} \cdots A_1$. Here, we consider that the control parameters are given as follows:

$$K_0 = \frac{\prod_{i=1}^{l} a_i}{1 - \prod_{i=1}^{l} a_i}, \tag{11.12}$$
$$K_1 = \frac{1}{a_l}, K_2 = \frac{1}{a_1}, K_3 = \frac{1}{a_2}, \dots, K_l = \frac{1}{a_{l-1}}.$$

Then we obtain

$$M = \frac{1}{1 - \prod_{i=1}^{l} a_i} \begin{pmatrix} \prod_{i=1}^{l} a_i & -a_l \prod_{i=1}^{l} a_i & 0 & \cdots & 0 \\ \prod_{i=1}^{l-1} a_i & -a_l \prod_{i=1}^{l-1} a_i & 0 & \cdots & 0 \\ \vdots & \vdots & \vdots & \ddots & \vdots \\ a_1 & -a_1 a_l & 0 & \cdots & 0 \\ 1 & -a_l & 0 & \cdots & 0 \end{pmatrix}. \tag{11.13}$$

The eigenvalues of the matrix M must be zero, and hence the system (11.6) can exhibit a superstable periodic orbit using the parameters (11.12).

Here we consider a typical example of the case of 4-periodic orbits embedded in the Bernoulli shift map for simplicity. In this case, the dynamics can be described as follows:

$$\begin{cases} x_0(n+1) = F(x_0(n) - K_0(x_4(n) - x_0(n))), \\ x_1(n+1) = x_0(n) - K_0(x_4(n) - x_0(n)), \\ x_2(n+1) = x_1(n) - K_0 K_1((x_4(n) - x_0(n)), \\ x_3(n+1) = x_2(n) - K_0 K_1^2((x_4(n) - x_0(n)), \\ x_4(n+1) = x_3(n) - K_0 K_1^3((x_4(n) - x_0(n)). \end{cases} \tag{11.14}$$

Here, the time-variant control gain $K(n)$, which depends on the slope of the original chaos map, is set to a constant gain K_1, because the slope of the Bernoulli shift map is constant everywhere ($a = a_1 = a_2 = \cdots = a_l$). If the feedback gain K_0 is equal to zero, this system can be identified with the original chaotic map F.

The stability of the periodic orbit can be analyzed by calculating the eigenvalues of the matrix product $M = A_4A_3A_2A_1$. And we set the control parameters as followings:

$$K_0 = \frac{a^4}{1-a^4}, K_1 = \frac{1}{a}. \tag{11.15}$$

Then we obtain

$$M = \frac{1}{1-a^4}\begin{pmatrix} a^4 & -a^5 & 0 & 0 & 0 \\ a^3 & -a^4 & 0 & 0 & 0 \\ a^2 & -a^3 & 0 & 0 & 0 \\ a & -a^2 & 0 & 0 & 0 \\ 1 & -a & 0 & 0 & 0 \end{pmatrix}. \tag{11.16}$$

The eigenvalues of the matrix (11.16) must be zero as with the matrix (11.13). Therefore, the system (11.14) exhibits superstable periodic orbits with period 4 on the parameters (11.15).

Figure 11.2 shows a time-domain waveform of superstable periodic orbits. We can observe that the trajectory converges to a periodic orbit within a finite time. The black area in Fig. 11.3 indicates a stability region in a parameter space of K_0 and K_1. In the previous method, the parameter space of K that satisfies (4) is $-\frac{17}{15} < K < -1$. Therefore, we can confirm that the parameter space of the proposed method is wider than that in the previous method.

11.4 Laboratory Results

Figure 11.4 shows a piecewise-constant chaos generator proposed in [4]. Here a' and f' represents linear and nonlinear amplifiers, respectively; τ_c is a time constant of integration. The dotted frame labeled S/H means a sample and hold block, which consists of a periodically operating switch and two voltage followers realized by operational amplifiers. The switch is closed for a moment every T[s]. The input voltage v is copied instantaneously to a capacitor at the time $t = nT$, and the output voltage v_n is held until $t = (n+1)T$. That is,

$$v_n = v(nT) \quad \text{for} \quad nT \leq t < (n+1)T, \tag{11.17}$$

where n is any positive integer and T is a switching period. Then the circuit dynamics are described by

$$\begin{aligned} v_{n+1} &= v_n - \frac{1}{\tau_c}\int_{nT}^{nT+T} (v_n - a'v_n - f'(v_n))dt \\ &= v_n - \frac{T}{\tau_c}(v_n - a'v_n - f'(v_n)). \end{aligned} \tag{11.18}$$

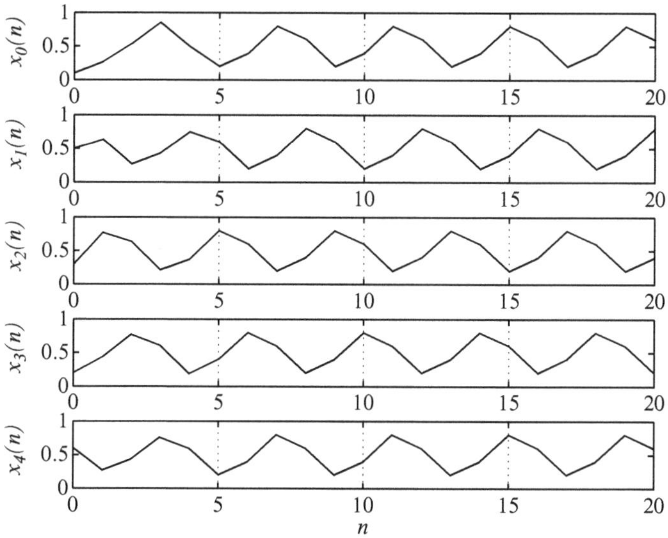

Fig. 11.2 Typical time-domain waveform shapes of the stability-transformed system ($l = 4$, $a = 2$, $K_0 = -\frac{16}{15}$, $K_1 = \frac{1}{2}$)

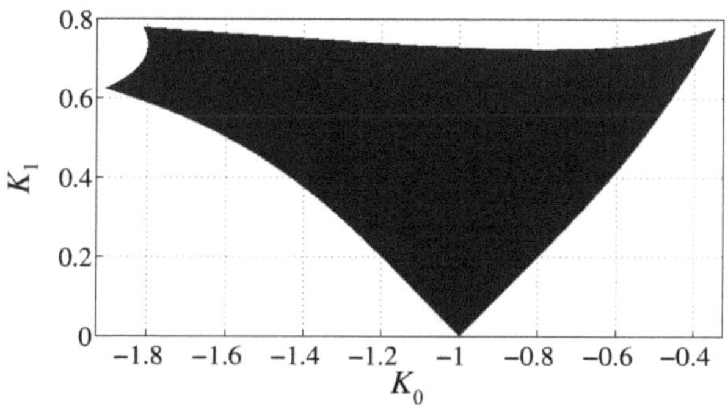

Fig. 11.3 Stability region in a parameter space of K_0 and K_1 ($l = 4$, $a = 2$)

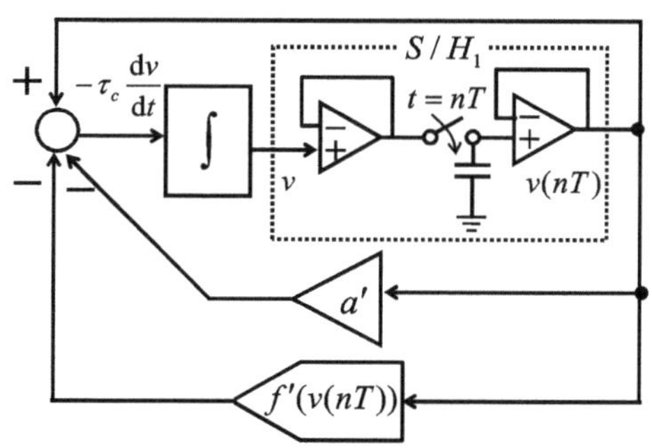

Fig. 11.4 Circuit diagram of the piecewise-constant system

Here we use the following normalized variables and dimensionless parameters:

$$x = \frac{v}{V}, \tau = \frac{1}{T}t, a = \frac{T}{\tau_c}a' \quad \text{and} \quad f(x) = \frac{1}{V}\frac{T}{\tau_c}f'(V \cdot x). \tag{11.19}$$

In this paper, we adjust the parameter T to be equal to rc for simplicity. Thereby the return map F from $x(n)$ to $x(n+1)$ can be described explicitly as follows:

$$F : x(n) \mapsto x(n+1), \quad x(n+1) = F(x(n)) = ax(n) + f(x(n)). \tag{11.20}$$

We consider the case in which the nonlinear function $f(x(n))$ is defined by

$$f(x(n)) = \begin{cases} 0 & \text{for } x(n) \leq \frac{1}{2}, \\ -\frac{a}{2} & \text{for } x(n) > \frac{1}{2}. \end{cases} \tag{11.21}$$

Consequently, we can derive a simple 1-D return map

$$F(x(n)) = \begin{cases} ax(n) & \text{for } x(n) \leq \frac{1}{2}, \\ a(x(n) - \frac{1}{2}) & \text{for } x(n) > \frac{1}{2}. \end{cases} \tag{11.22}$$

In this case, the return map F is the Bernoulli shift map, as shown in Fig. 11.5. Figure 11.5b shows an example of time evolutions of the discrete dynamics (11.22) corresponding to the time series as shown in Fig. 11.5a. Adjusting the gain factor a and the nonlinear function $f(x(n))$, we can construct a continuous chaos generator corresponding to any 1-D return map $F(x(n))$.

A block diagram of the system (11.14) is shown in Fig. 11.6. In this figure, the dotted frame labeled F corresponds to the chaos generator shown in Fig. 11.4. We indicate the relationship between $x_0(n)$ and $x_0(n+1)$ obtained from calculation and the laboratory circuit. Figure 11.7a, b shows the original chaos system (11.22) and the transformed system (11.14), respectively. Figure 11.8a shows the original chaos system, which is generated by the piecewise-constant chaos generator in Fig. 11.4, and Fig. 11.8b shows the transformed system (11.14) with stable 4-periodic orbit. We can confirm that the proposed system can be realized with the real system.

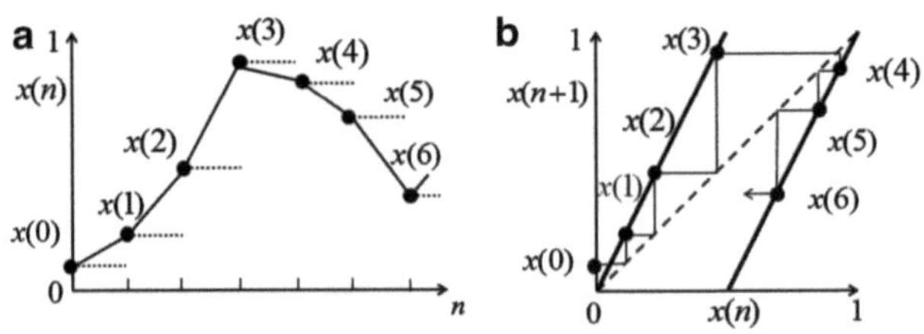

Fig. 11.5 An example of (**a**) a time-domain waveform and corresponding (**b**) a return map

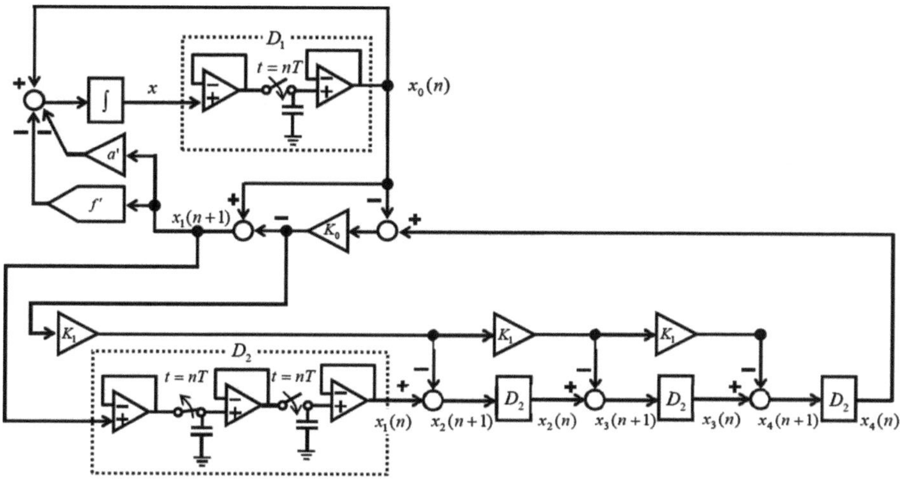

Fig. 11.6 Stability transformation system based on the piecewise-constant system

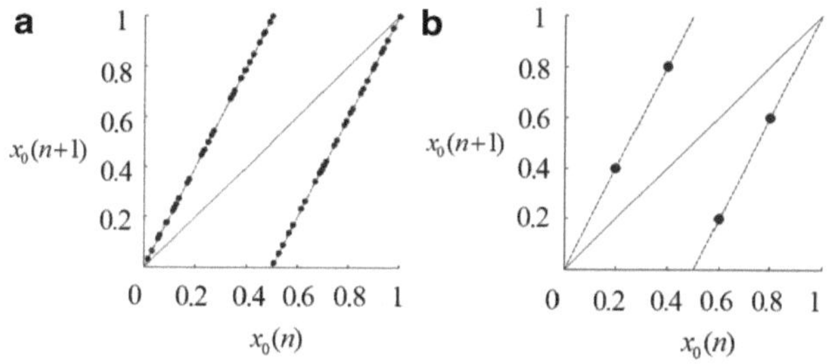

Fig. 11.7 Relationship between $x_0(n)$ and $x_0(n+1)$, (**a**) original chaos system ($a = 2$), (**b**) transformed system

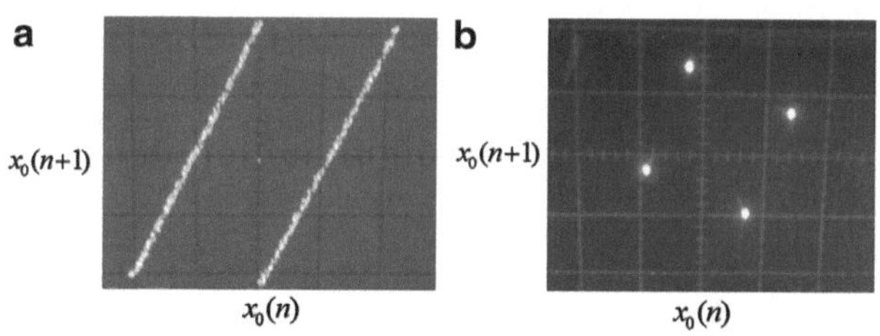

Fig. 11.8 Laboratory measurements, (**a**) original chaos system, (**b**) four-periodic orbit obtained from the stability transformed system, *vertical* axis: 1[V/div] and *horizontal* axis: 1[V/div]

11.5 Conclusion

In this paper, we have proposed a stability transformation method to generate stabilized UPOs that are embedded in a chaos attractor of a piecewise-constant chaotic system. The transformed system becomes superstable by the use of parameters determined from the derivatives at periodic points. Also, we analyzed a stability region in a parameter space of control gain and confirmed some results by laboratory measurements.

A future problem is to generalize the system. Now we try to analyze a domain of attraction in the transformed system.

References

1. Ott, E., Grebogi, C., Yorke, J.A.: Controlling chaos. Phys. Rev. Lett. **64**(11), 1196–1199 (1990)
2. Pyragas, K.: Continuous control of chaos by self-controlling feed-back. Phys. Lett. A **170**, 421–428 (1992)
3. Chen, G., Yu, X.: On time-delayed feedback control of chaotic systems. IEEE Trans. Circuits Syst. I **46**(6), 767–772 (1999)
4. Pyragas, K.: Control of chaos via extended delay feedback. Phys. Lett. A **206**, 323–330 (1995)
5. Kokame, H., Hirata, K., Konishi, K., Mori, T.: Difference feedback can stabilize uncertain steady states. IEEE Trans. Autom. Contr. **46**, 1908–1913 (2001)
6. Ushio, T.: Limitation of delayed feedback control in nonlinear discrete-time systems. IEEE Trans. Circuits Syst. I Fundam. Theory Appl. **43**(9), 815–816 (1996)
7. Nakajima, H.: On analytical properties of delayed feedback control of chaos. Phy. Lett. A **232**(28), 207–210 (1997)
8. Konishi, K., Kokame, H.: Observer-based delayed-feedback control for time-discrete chaos systems. Phys. Lett. A **248**, 359–368 (1998)
9. Yamamoto, S., Hino, T., Ushio, T.: Dynamic delayed feedback controllers for chaotic discrete-time systems. IEEE Trans. CAS-I **46**(6), 767–772 (2001)
10. Ushio, T., Yamamoto, S.: Prediction-based control of chaos. Phys. Lett. A **264**, 30–35 (1999)
11. Pyragas, K., Pyragas, V., Kiss, I.Z., Hudson, J.L.: Adaptive control of unknown unstable steady states of dynamical systems. Phys. Rev. E **70**, 026215 (2004)
12. Jiang, G.-P., Chen, G., Tang, W.K.-S.: Stabilizing unstable equilibria of chaotic systems from a state observer approach. IEEE Trans. Circuits Syst. II Expr. Briefs **51**(6), 281–288, (2004)
13. Hassouneh, M.A., Lee, H.-C., Abed, E.H.: Washout filters in feedback control: Benefits, limitations and extensions. In: Proceedings of the American Control Conference, vol. 5, pp. 3950–3955 (2004)
14. Huijberts, H.: Linear controllers for the stabilization of unknown steady states of chaotic systems. IEEE Trans. Circuits Syst. I **53**(10), 2246–2254 (2006)
15. Pingel, D., Schmelcher, P., Diakonos, F.K.: Detecting unstable periodic orbits of chaotic dynamical systems. Phys. Rev. Lett. **78**(25), 4733–4736 (1997)
16. Pingel, D., Schmelcher, P., Diakonos, F.K.: Stability transformation: a tool to solve nonlinear problems. Phys. Rep. **400**, 67–148 (2004)

17. Tsubone, T., Kambayashi, N.: Chaos and periodic oscillation of a multirate sampled data dynamical system. IEICE Ttrans. Fundamentals **E90-A**(9), 1924–1931 (2007)
18. Tsubone, T., Wada, Y.: Stabilizing unknown periodic orbits of a chaotic spiking oscillator. IEICE Ttrans. Fundamentals **E92-A**(5), 1316–1321 (2009)
19. Callegori, S.: Some more robustness conditions for the invariant density of a class of 1D maps under additive noise, circuit theory and design. In: 18th European Conference on ECCTD, pp. 1038–1041 (2007)

Chapter 12
Lorenz Attractors and Generalizations: Geometric and Topological Aspects

Natalia Klinshpont

Abstract In this paper, the problem of topological distinguishing of Lorenz-like attractors and their generalizations is discussed. New cardinal-valued topological invariants for Lorenz-type attractors and generalizations are constructed. A generalization is considered of Williams's well-known model of the attractor in the Lorenz system, the inverse limit of semiflows on branched manifolds that are suspensions over a discontinuous expanding map of a closed linear interval. The generalization consists in the consideration of maps with several discontinuity points, rather than one. A cardinal-valued topological invariant L-manuscript is constructed that distinguishes a continuum of nonhomeomorphic generalized models. A topological invariant distinguishing a continuum of nonhomeomorphic geometric Lorenz attractors is obtained as a consequence. An analogous cardinal-valued invariant is constructed for attractors of Lorenz-type maps. The kneading invariant is not a topological invariant distinguishing attractors as sets; there exists an uncountable set of mutually nonconjugating Lorenz-type maps having homeomorphic attractors.

12.1 Introduction

The history of Lorenz system investigation began in 1963 and led the scientific community to the thesis that it was impossible to make a long-term forecast of dynamical systems with a complex orbital behavior (such as in meteorology) as a result of the so-called butterfly effect, whereby small changes in the system (in particular, in the initial data) lead to large changes in its behavior over relatively moderate time intervals (in meteorology about 30 days).

N. Klinshpont (✉)
Obninsk Institute for Nuclear Power Engineering, 249040, Studgorodok 1,
Obninsk, Kaluga Region, Russia
e-mail: klinshpont-n@yandex.ru

C. Grácio et al. (eds.), *Nonlinear Maps and their Applications*, Springer Proceedings in Mathematics & Statistics 57, DOI 10.1007/978-1-4614-9161-3_12,

It turns out that all relevant information about such a system focuses on attracting sets—attractors, which can have a very complicated geometric structure (so-called strange attractors).

Hence, this was not just a problem of geometric research, but also a paradigm for the statistical study of the dynamics of strange attractors.

As Romen Plykin wrote [10]:

> The problem of the topological classification of manifolds, moreover of general sets, was long ago abandoned by mathematicians not only due to its algorithmic insolvability, but also due to the extreme kaleidoscopicity and complexity of objects being a stumbling block on the way to the conceptual integrity of a research program. However, for the maps, sets and manifolds arising from the problems of the theory of dynamical systems, the classification problems sometimes may be reasonably solved.

The topological classification problem has been solved for expanding hyperbolic attractors of codimension one [9] and for quasiattractors of the Smale horseshoe [2, 14].

This article presents some results obtained by on the topological classification problem of Lorenz-like attractors and their generalizations—a survey of recent results on generalizations of geometric Lorenz attractors (Theorems 4 and 5), and some new results on attractors of Lorenz-type maps (Theorems 2 and 3).

The author is grateful to R.V. Plykin and E.A. Sataev for fruitful discussions.

12.2 Lorenz Attractor and Geometrical Lorenz Flow

In 1963, Lorenz [7] carried out numerical research of the system

$$\dot{x}_1 = -\sigma x_1 + \sigma x_2,\ \dot{x}_2 = \rho x_1 - x_2 - x_1 x_3,\ \dot{x}_3 = -\beta x_3 + x_1 x_2, \tag{12.1}$$

with $\sigma = 10$, $\beta = \frac{8}{3}$, $\rho = 28$.

Almost all points in the phase space are attracted by a strange attractor. The attention of experts in dynamical systems was drawn to the Lorenz system after the papers [8, 17], in which the following properties of the system were distinguished:

(L1) There exists a bounded domain U absorbing all trajectories of the system.

(L2) The point $O(0,0,0)$ is fixed. The linearized system in the neighborhood of O has real eigenvalues such that $\lambda_1 < \lambda_2 < 0 < \lambda_3$ and $\lambda_2 + \lambda_3 > 0$. Hence the point O has a two-dimensional stable manifold $W^s(O)$ and a one-dimensional unstable manifold $W^u(O)$ consisting of two trajectories.

In Fig. 12.1 we depict the behavior of solutions of the Lorenz system as obtained from numerical experiments. The figure is borrowed from the survey [13].

Afraimovich et al. [1] and Guckenheimer and Williams [3] introduced two geometric models reflecting the behavior of solutions of the Lorenz system for parameters close to the classical values that follow from numerical calculations. We shall not describe these models in detail, but postulate instead the properties of the flow Φ_t by adding two further properties to (L1) and (L2). In what follows, we call Φ_t the *geometric Lorenz flow*.

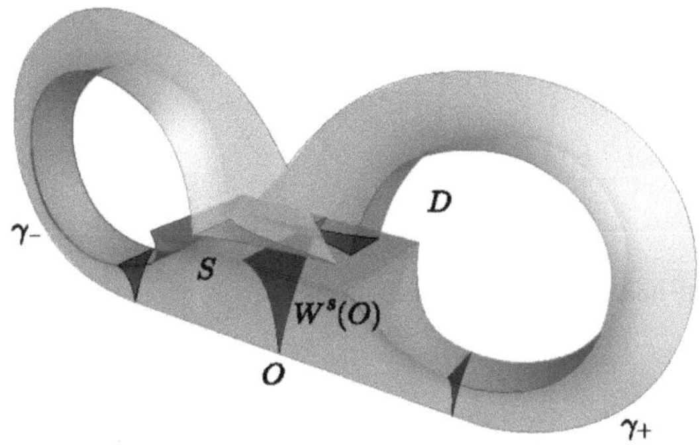

Fig. 12.1 Lorenz flow Φ_t

(L3) There exists a two-dimensional platform D transversal to the trajectories of the flow; $W^s(O)$ intersects D by a curve S. The first return map $T : D \setminus S \to D$ is schematically represented in Fig. 12.2. There exist limits $P_1 = \lim_{y \to 0^-} T(x,y)$ and $P_2 = \lim_{y \to 0^+} T(x,y)$. One can introduce a coordinate system on D such that $D = [0,1] \times [0,1]$. The line S is defined by the equation $y = 0$. There exist constants $\theta_s > 0$, $\theta_u > 0$, $\lambda \in (0,1)$, and $q > 1$ such that $\theta_s \theta_u < 1$ and:

- If l is the graph of a function $x = h(y)$, $|h'(y)| < \theta_u$, and $l \cap S = \emptyset$, then $T(l)$ is a similar curve. That is, $T(l)$ is the graph of a function $x = H(y)$, $|H'(y)| < \theta_u$; if p and p' are points in l, then $d(Tp, Tp') > qd(p, p')$ (d is the distance between the points).
- If w is the graph of a function $y = g(x)$, $|g'(x)| < \theta_s$, then $T^{-1}(w \cap T(D_i))$ is the graph of a function $y = G(x)$ and $|G'(x)| < \theta_s$. If p and p' are points on the curve $w \cap T(D_i)$, then $d(T^{-1}p, T^{-1}p') > d(p, p')/\lambda$.

As a consequence, we have that in the domain D there exists a stable foliation invariant under T [1]. Factorization along the leaves of the stable foliation transforms the domain D so that it is factorized onto some interval $[0,1] = I$. The map T is projected to a map $f : I \to I$ of the interval. Let s be the projection of the line S onto I. The map f is monotonic on the intervals $[0, s)$ and $(s, 1]$, $\lim_{x \to s^-} f(x) = p_1$ and $\lim_{x \to s^+} f(x) = p_2$ (see Fig. 12.2). In [3, 15, 16], the authors impose on f an additional condition:

(L4) The map $f|_{[p_1, p_2]}$ is locally eventually onto, that is, for each subinterval $J \subset [p_1, p_2]$, there exists $n > 0$ such that $f^n(J) = [p_1, p_2]$.

The map T satisfying property (L3) is called a *Lorenz-type map*.

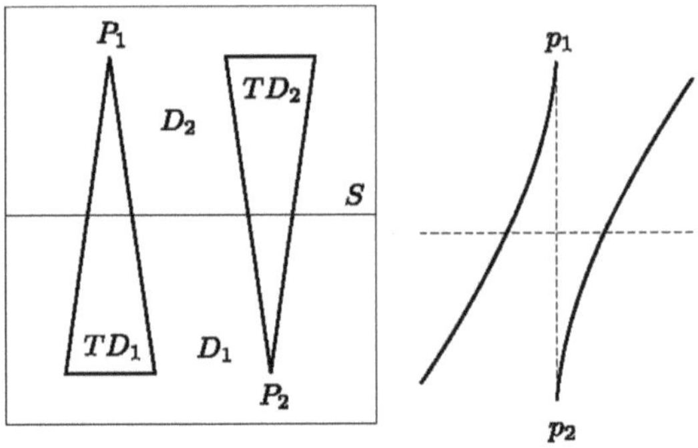

Fig. 12.2 Lorenz map and corresponding map of interval

For the actual Lorenz system, properties (L3) and (L4) have been repeatedly numerically verified by many authors. A proof of a weak version of properties (L3) and (L4) has recently been carried out by Tucker [12], who combined analytical methods and numerical calculations.

The *attractor of the flow* Φ_t is the set $\Lambda = \bigcap_{t \geqslant 0} \Phi_t(U)$. The *attractor of the Lorenz-type map* T is the set $\Sigma_T = clos \bigcap_{n \geqslant 0} T^n(D)$. The set Σ_T is a cross-section of the set Λ. If conditions (L1)–(L4) are satisfied, then periodic trajectories are dense in Λ and the flow Φ_t is topologically transitive [1, 3, 15].

12.3 On the Problem of Homeomorphism of Attractors of Lorenz-Type Maps

We emphasize that the problem of homeomorphism of attractors means the topological equivalence of attractors as sets, taking no account of the dynamics on these attractors. The geometric structure of attractors Σ_T has been analyzed in [4,5], where a topological invariant distinguishing a continuum of nonhomeomorphic attractors is constructed. The set Σ_T has a very complicated cellular geometric structure, caused by the presence of a countable set of curvilinear triangles or digons (lunes). Moreover, for each point $x = T^k(P_i)$, there exists a subset of Σ_T homeomorphic to a cone over the Cantor set with vertex at the point x.

Consider any point $z \in \Sigma_T \setminus \bigcup_{k \geq 0} T^k(P_1 \cup P_2)$. For all $k > 0$ $T^{-k}z \in \Sigma_T$, $T^{-k}z \notin S$. Define the sequence $\alpha_T(z) = \alpha_1, \alpha_2, \ldots$, where $\alpha_k = 1$ if $T^{-k}z$ lies above S and $\alpha_k = 0$ if $T^{-k}z$ lies below S. We call this sequence the *symbolic prehistory* of the point z.

Consider the set

$$A_T = clos \bigcup_{k \geq 0} T^k(P_1 \cup P_2) \setminus \bigcup_{k \geq 0} T^k(P_1 \cup P_2)$$

and let Ω_T be the set of symbolic prehistories of points in A_T:

$$\Omega_T = \{\omega : \exists z \in A_T, \alpha_T(z) = \omega\}.$$

For each $\omega \in \Omega_T$, introduce the set $C_T(\omega) = \{z \in A_T : \alpha_T(z) = \omega\}$. Let $card(C_T(\omega))$ be the cardinal number of $C_T(\omega)$. Set

$$\chi_T = \{card(C_T(\omega)) : \omega \in \Omega_T\}.$$

Theorem 1. *Let T_1 and T_2 be Lorenz-type maps. If the set Σ_{T_1} is homeomorphic to the set Σ_{T_2}, then $\chi_{T_1} = \chi_{T_2}$. The set χ_T has the following properties:*

- *If $n \in \chi_T$ (n a natural number), then $1 \in \chi_T$.*
- *If $n \in \chi_T$, then there exist numbers $n_1 \in \chi_T$ and $n_2 \in \chi_T$ such that $n = n_1 + n_2$ or $n = n_1 + n_2 + 1$.*

For every increasing sequence of natural numbers with the properties

1. *$n_1 = 1$;*
2. *for all $i \geq 2$ there are $k, l \leq i$ such that $n_i = n_k + n_l$ or $n_i = n_k + n_l + 1$,*

there exists a Lorenz-type map T such that $\chi_T = \{n_1, n_2, n_3, \dots, n_k, \dots, \aleph_0, c\}$.

Theorem 1 is proved in [5]. The geometric sense of the invariant χ_T is as follows. Each point z of the set

$$\Sigma_T \setminus \bigcup_{k \geq 0} T^k(P_1 \cup P_2)$$

lies on a curve l_z (so-called unstable layer) belonging to Σ_T. We call points of A_T *marked points*. The invariant χ_T is the system of cardinalities of marked points lying on unstable layers. The arithmetic properties of χ_T are illustrated in Fig. 12.3.

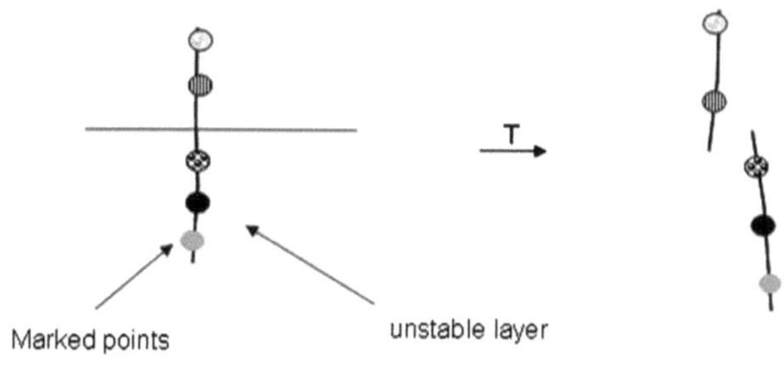

Fig. 12.3 Arithmetic properties of χ_T

Remarks

1. By Theorem 1, there exists an uncountable set of nonhomeomorphic attractors.
2. The condition $\chi_{T_1} = \chi_{T_2}$ is not sufficient for topological equivalence of Σ_{T_1} and Σ_{T_2}. Moreover, there exists an uncountable set consisting of nonhomeomorphic attractors $\Sigma_{T\alpha}$ having the same sets $\chi_{T\alpha}$.

3. Every type of nonhomeomorphic attractor can be obtained by a small C^1 perturbation of the map of the interval f, generating an attractor.
4. The kneading invariant of topological conjugation of Lorenz-type maps is not a topological invariant distinguishing attractors as sets.
   The *kneading sequences* of the Lorenz map of the interval $f\colon [p_1,p_2] \to [p_1,p_2]$ are the sequences $l=l_1,l_2,\ldots,l_n,\ldots$ and $r=r_1,r_2,\ldots,r_n,\ldots$ such that
   $l_n = -1$ if $f^n(p_1) < s$, $l_n = 0$ if $f^n(p_1) = s$, and $l_n = 1$ if $f^n(p_1) > s$,
   $r_n = -1$ if $f^n(p_2) < s$, $r_n = 0$ if $f^n(p_2) = s$, and $r_n = 1$ if $f^n(p_2) > s$,
   where s is a point of discontinuity.
   It is known [11] that the set $\{l,r,-l,-r\}$ is an invariant of the topological conjugacy of Lorenz-type maps. The following theorem shows that the kneading invariant is not a topological invariant distinguishing attractors as sets of points.

   Theorem 2. *There exists an uncountable set of mutually nonconjugate Lorenz-type maps T_c having homeomorphic sets Σ_{T_c}.*

   The proof is the construction of an uncountable set of nonconjugate Lorenz-type maps of the interval $f_c\colon [p_1,p_2] \to [p_1,p_2]$ having the same set B of limiting points of trajectories $f_c^n(p_k)$. This set B is a Cantor set, and the maps $f_c|_B$ restricted to B are equal.
5. One can consider more generalized Lorenz-type maps with several lines of discontinuity (see Fig. 12.4). An analogous topological invariant is constructed for the general case, but the arithmetic properties are more complicated.

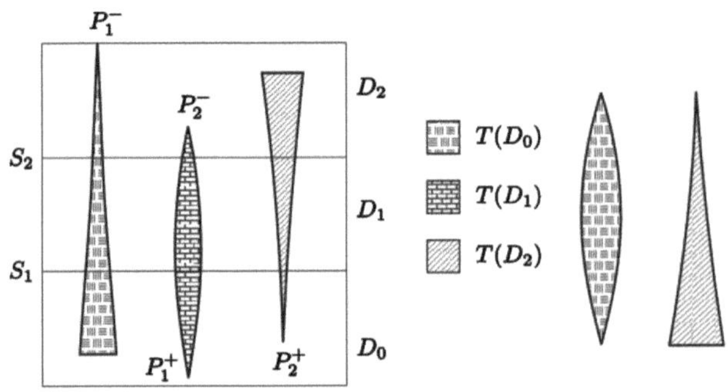

Fig. 12.4 Lorenz map with two lines of discontinuity

Theorem 3. *Let T be a Lorenz-type map with k lines of discontinuity. The set χ_T has the following properties:*

- *If $n \in \chi_T$ (n a natural number), then $1 \in \chi_T$.*
- *If $n \in \chi_T$, then there exist numbers $n_1, n_2, \ldots, n_m \in \chi_T$ and $0 \le i \le k$ such that $n = n_1 + n_2 + n_3 + \ldots + n_m + i$.*

For every increasing sequence of natural numbers with the properties

(a) $n_1 = 1$,
(b) for all $j \geq 2$ *there are numbers* $n_1, n_2, \ldots, n_m \in \chi_T (m \geq 1)$ *such that* $n_j = n_1 + n_2 + \ldots + n_m$.

there exists a Lorenz-type map T such that $\chi_T = \{n_1, n_2, n_3, \ldots, n_k, \ldots, \aleph_0, c\}$.

12.4 On the Problem of Homeomorphism of Geometric Lorenz Attractors and Their Generalizations

Williams [15, 16] introduced the following model of the Lorenz system attractor. He assumed that there exists a one-dimensional invariant stable foliation in the area U transverse to trajectories of the flow Φ_t. The result of the area U factorization on this foliation has in general a very complicated structure. In addition, Williams assumed that there is a neighborhood U_1 of the attractor Λ such that U_1 factorized by the stable foliation is the branched manifold (we denote it by L), and Φ_t is projected to the semiflow ϕ_t, $t > 0$, on L (see Fig. 12.5 left).

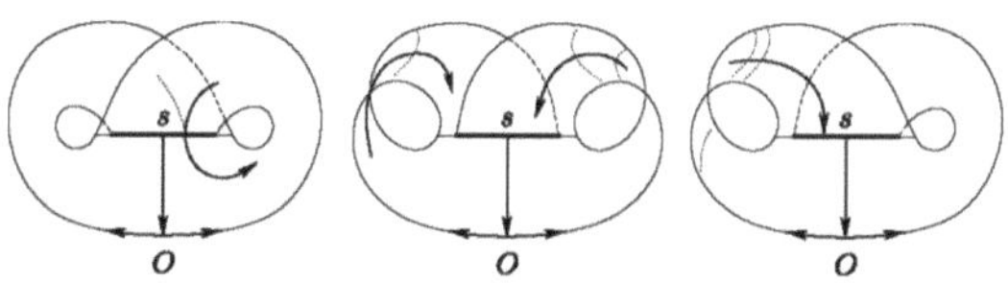

Fig. 12.5 Branched manifolds of three kinds: orientable, nonorientable and semiorientable

Let L be a smooth branched manifold of one of three kinds: orientable, semiorientable, or nonorientable (see Fig. 12.5). The horizontal interval I contains a line of branching, which is indicated by a heavy line. The fixed point O of the flow Φ_t is projected to the saddle fixed point of the semiflow ϕ_t (we also denote this point by O). Let $W^u(O)$ and $W^s(O)$ be respectively unstable and stable manifolds of the point O (considered as the fixed point of the semiflow ϕ_t).

The maps $\phi_t : L \to L, t \in \mathbb{R}^+$ form an inverse spectrum of topological spaces. The corresponding inverse limit $\hat{L} = (\varprojlim L, \phi_t, t \in \mathbb{R}^+)$ consists of points $\hat{x} = (x_\theta)_{\theta \geq 0}$, where $\phi_t(x_\theta) = x_{\theta - t}$.

On $\hat{L}$, the flow $\hat{\phi}_t$ is defined as follows: if $t > 0$, then $\hat{\phi}_t(\hat{x}) = (y_\theta)_{\theta \geq 0}$, where $y_\theta = \phi_t x_\theta$; if $t < 0$, then $\hat{\phi}_t(\hat{x}) = (y_\theta)_{\theta \geq 0}$, where $y_\theta = x_{\theta - t}$.

The pair $(\hat{L}, \hat{\phi}_t)$ is called a *geometric Lorenz attractor*. Such a representation is due to Williams [15,16], who investigated the geometric structure of $\hat{L}$. As the main tool, Williams used kneading sequences: pairs of sequences of zeros and ones characterizing the order in which the left-hand and right-hand unstable manifolds of the point O go around the right-hand and left-hand "holes" in the branched manifold L. As Williams has shown [16], if two sets $\hat{L}_1$ and $\hat{L}_2$ are homeomorphic and their homeomorphism is close to the identity in a certain sense, then the corresponding pairs of kneading sequences coincide. Thus, the kneading sequences are a conditional topological invariant (the homeomorphism must be close to the identity).

The problem of the construction of an absolute topological invariant of Lorenz geometric attractors is solved in [6], where a cardinal-valued topological invariant L-manuscript is constructed. In [6], a generalization is considered of Williams's model to the case of branched manifolds carrying semiflows with several fixed hyperbolic points. Such semiflows are suspensions over discontinuous expanding maps of an interval with several points of discontinuity.

12.4.1 Generalization of Williams's Model

Let f be a map of the interval $[0,1]$ with $n \geq 1$ discontinuity points s_1, ..., s_n that is monotonic and differentiable on each interval $\Delta_1 = [0,s_1), \Delta_2 = (s_1,s_2)$, ..., $\Delta_{n+1} = (s_n,1]$. There obviously exist limits $\lim_{x\to s_i-0} f(x) = p_i^-$ and $\lim_{x\to s_i+0} f(x) = p_i^+$ [see Fig. 12.7 (left)]. We additionally assume that the map f has the following properties:

1. If $n = 1$, then $|f'| > \sqrt{2}$.
2. If $n \geq 2$, then $|f'| > 2$, and in addition, for each Δ_i, there exists k_i such that $\bigcup_{m=0}^{k_i} f^m(\Delta_i) = (0,1)$.

It follows from conditions (1) and (2) that the set of all preimages of each point is dense in [0,1], the periodic points are dense, f is topologically transitive.

Now for a fixed map f we construct a branched manifold carrying a semiflow. Consider the square $\Pi = [0,1] \times [0,1]$ cut along the lines $s_k \times [0.5,1]$ (see Fig. 12.6). We denote points on the left and the right edges by $s_k^- \times x$ and $s_k^+ \times x$, respectively

On Π, we define a vertical flow with fixed points $O_k = s_k \times 0.5$, $k \in 1,2,\ldots,n$, as follows:

$$\dot{x} = 0, \dot{y} = g(x,y),$$

where $g(x,y)$ is a continuous function, $g(x,y) = 1$ outside neighborhoods of the points O_k, and $g(x,y)$ takes positive values outside the points O_k at which $g = 0$.

We now identify points on the upper and lower parts of the boundary of the cut square Π: $x \times 1 \sim f(x) \times 0$ if x is not a discontinuity point; $s_k^- \times 1 \sim p_k^- \times 0$; $s_k^+ \times 1 \sim p_k^+ \times 0$. We call the result the branched 2-manifold L [see Fig. 12.7 (right)]. The flow on the square Π induces a semiflow φ_t, $t \geq 0$, on L. This semiflow φ_t is not a flow, since L contains the line $[0,1] \times 0$ containing the set of branch points resulting from the points glued together. The semiflow φ_t has n fixed saddle-like points (in the topological sense) O_k, $k = 1,\ldots,n$.

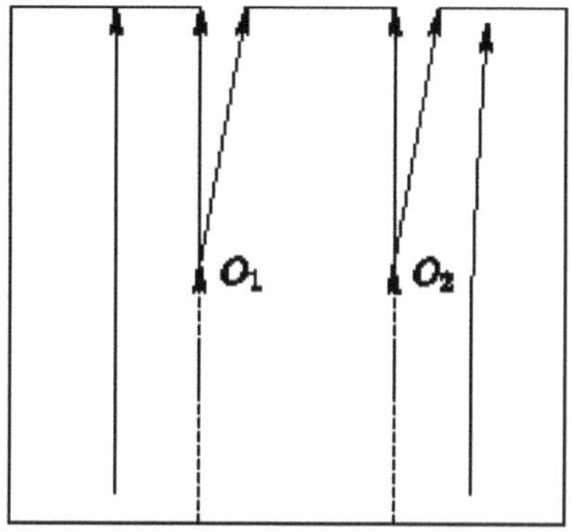

Fig. 12.6 A cut square Π with a vertical flow

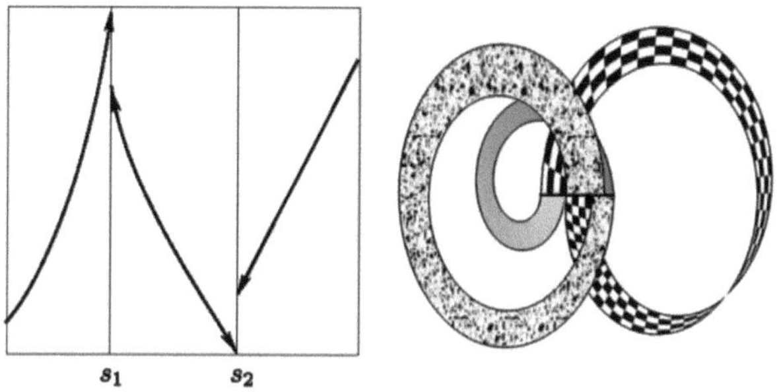

Fig. 12.7 The map of an interval with two discontinuity points and the corresponding branched manifold

We set $\hat{L} = (\varprojlim L, \varphi_t, t \in \mathbb{R}^+)$.

Definition 1. We call the pair $(\hat{L}, \hat{\varphi}_t)$ a *generalized geometric Lorenz attractor.*

Slightly abusing the terminology, we shall mean by a geometric attractor the space $\hat{L}$ itself, implying that the flow $\hat{\varphi}_t$ on it is fixed. For $n = 1$, the pair $(\hat{L}, \hat{\varphi}_t)$ coincides with geometric Lorenz attractors (in the sense of Williams).

12.4.2 Construction of a Topological Invariant

Now we shall construct a topological invariant for the attractors $\hat{L}$ regarded as sets, taking no account of the flow dynamics on them. An indication of the complicated geometric structure of an attractor $\hat{L}$ is the presence of points with neighborhoods containing the Cartesian product of an interval and a cone over the Cantor set (the so-called Cantor book). Let $\hat{O}_k$ be a fixed point of the flow $\hat{\varphi}_t$: $\hat{O}_k = (x_\theta)_{\theta > 0}$, where $x_\theta = O_k$ for all $\theta \geq 0$. We denote by $\hat{W}^u(\hat{O}_k)$ the unstable manifold of the point $\hat{O}_k$; $\hat{W}^u(\hat{O}_k)$ consists of points $\hat{z} = (z_\theta)_{\theta > 0}$ such that $z_\theta \to O_k$ as $\theta \to \infty$.

Proposition 1. *An arbitrary point $\hat{z} \in \hat{L}$ lies on the spine of a Cantor book embedded in $\hat{L}$ if and only if $\hat{z} \in \bigcup_{k=1}^{n} \hat{W}^u(\hat{O}_k)$.*

This proposition was established by Williams in a particular case considered in [15, 16]. Let $\hat{z} \in \hat{L} \setminus \bigcup_{k=1}^{n} \hat{W}^u(\hat{O}_k)$. The corresponding negative semitrajectory $\hat{z} = (z_\theta)_{\theta \geq 0}$ crosses the line $I = [0,1] \times \{0\}$ infinitely many times. Denote by $x_1, x_2, x_3, \ldots$ the consecutive points of crossing.

Let $\alpha(\hat{z}) = \alpha_1, \alpha_2, \alpha_3, \ldots$ be the sequence of symbols $\{0,1,\ldots,n\}$ formed as follows: $\alpha_i = k-1$ if $x_i \in \Delta_k$. We call $\alpha(\hat{z})$ the symbolic prehistory of the point $\hat{z}$. We shall say that sequences α and β are equivalent ($\alpha \sim \beta$) if they have the same "tails," that is, if there exist integers m and i such that $\alpha_{i+k} = \beta_{m+k}$ for $k > 0$.

Proposition 2.

1. *Points $\hat{z}_1$ and $\hat{z}_2$ lie in the same linearly connected component of the set $\hat{L} \setminus \bigcup_{k=1}^{n} \hat{W}^u(\hat{O}_k)$ if and only if $\alpha(\hat{z}_1) \sim \alpha(\hat{z}_2)$.*
2. *Each linearly connected component of $\hat{L} \setminus \bigcup_{k=1}^{n} \hat{W}^u(\hat{O}_k)$ has dimension 2 or 1. A one-dimensional component is the full trajectory of a point.*
3. *The two-dimensional components are dense in $\hat{L}$.*
4. *The set of two-dimensional linearly connected components of $\hat{L} \setminus \bigcup_{k=1}^{n} \hat{W}^u(\hat{O}_k)$ has the cardinality of the continuum.*

Denote by A_i the set of points $\hat{z} \in \hat{L} \setminus \bigcup_{k=1}^{n} \hat{W}^u(\hat{O}_k)$ that are limit points of $\hat{W}^u(\hat{O}_i)$, $i \in \{1,2,\ldots,n\}$:

$$A_i = clos \hat{W}^u(\hat{O}_i) \setminus \bigcup_{k=1}^{n} \hat{W}^u(\hat{O}_k).$$

Let Ω be the set of symbolic prehistories of points in $\bigcup_{k=1}^{n} A_k$:

$$\Omega = \{\alpha(\hat{z}) | \hat{z} \in \bigcup_{k=1}^{n} A_k\}.$$

Let $\hat{\Omega} = \Omega / \sim$ be the set of equivalence classes corresponding to the above-defined equivalence relation $\sim$. There exists a one-to-one correspondence between $\hat{\Omega}$ and the set of linearly connected components of $\hat{L} \setminus \bigcup_{k=1}^{n} \hat{W}^u(\hat{O}_k)$ containing points in $\bigcup_{k=1}^{n} A_k$. For this reason, we shall use for a linearly connected component the same notation as for the corresponding element of $\hat{\Omega}$. We observe that each point $\hat{z} \in \bigcup_{k=1}^{n} A_k$ lies with its full trajectory in some linearly connected component. Let $\{\hat{\varphi}_t(\hat{z})\}_{t \in R} = l(\hat{z})$ be this trajectory. We shall call the full trajectories of points $\hat{z} \in \bigcup_{k=1}^{n} A_k$ *marked trajectories*.

For each $\hat{\beta} \in \hat{\Omega}$, let $C_k(\hat{\beta})$ be the set of marked trajectories in A_k lying in the connected component $\hat{\beta}$:

$$C_k(\hat{\beta}) = \{l(\hat{z}) | \hat{z} \in A_k, \alpha(\hat{z}) \in \hat{\beta}\}.$$

In what follows, the notation $(\cdots)$ will mean an ordered set, while $\{\ldots\}$ means an unordered set.

Definition 2. We call the set $\chi = \chi(\hat{L}, \hat{\varphi}) = \{(cardC_k(\hat{\beta}))_{k=1,\ldots,n} | \hat{\beta} \in \hat{\Omega}\}$ the *L-manuscript.*

Theorem 4. *The L-manuscript is a homeomorphism invariant of generalized geometric Lorenz attractors. More precisely, if $\hat{L}_1$ and $\hat{L}_2$ are homeomorphic, then $\chi_1 = \chi_2$ up to a rearrangement: there exists a rearrangement $(i_1, i_2, \ldots, i_n)$ such that*

$$\chi_1 = \{(cardC_k(\hat{\beta}))_{k=1,\ldots,n} | \hat{\beta} \in \hat{\Omega}_1\} = \chi_2 = \{(cardC_{i_k}(\hat{\beta}))_{k=1,\ldots,n} | \hat{\beta} \in \hat{\Omega}_2\}.$$

Theorem 5. *Let $\Theta = \{\aleph_0, c, 0, 1, 2, 3, \ldots, n, \ldots\}$ be a system of cardinal numbers. For each $Y \subset \Theta \times \Theta \times \ldots \times \Theta$ containing $(c, c, \ldots, c)$ and not containing $(0, 0, \ldots, 0)$, there exists a pair $(\hat{L}, \hat{\varphi}_t)$ such that $\chi = Y$.*

See [6] for a proof.

The geometric sense of the *L*-manuscript invariant is as follows. The attractor is similar to a "book" with *n* spines (unstable manifolds of fixed points) and an uncountable set of "pages," which are linearly connected components of the set $\hat{L} \setminus \bigcup_{k=1}^{n} \hat{W}^u(\hat{O}_k)$. Some pages have marked trajectories (Fig. 12.8). The invariant *L*-manuscript is the system of cardinalities of marked trajectories lying on pages.

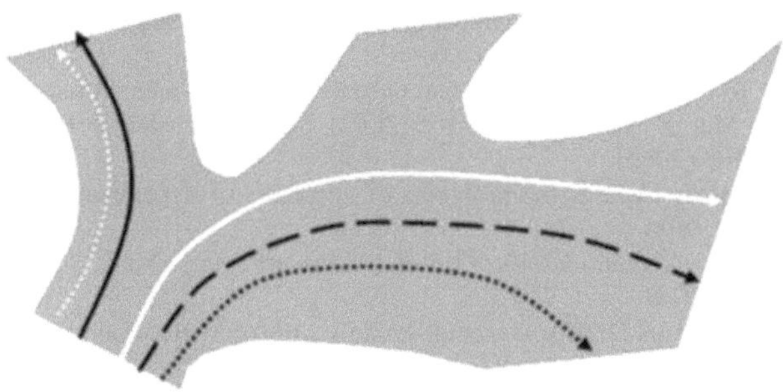

Fig. 12.8 Unstable layer with marked trajectories

Remarks

1. By Theorem 5, there exists a continuum of nonhomeomorphic generalized geometric Lorenz attractors.
2. The set $\chi = \chi(\hat{L}, \hat{\varphi})$ is a topological invariant of geometric Lorenz attractors, and for each system of distinct positive integers $n_1, n_2, n_3, \ldots$, there exists a pair $(\hat{L}, \hat{\varphi})$ such that $\chi(\hat{L}, \hat{\varphi}) = \{\aleph_0, c, n_1, n_2, n_3, \ldots\}$. There exists a continuum of nonhomeomorphic geometric Lorenz attractors.
3. All types of topologically distinct attractors can be obtained by a small C^1 perturbation of the map of the interval, generating an attractor.
4. The *L*-manuscript invariant is not complete. One can construct examples of a continuum of nonhomeomorphic attractors with the same value of *L*-manuscript.

Problem. Similarly, a series of cardinal valued topological invariants for Lorenz-type attractors can be constructed, but the problem of finding a complete invariant remains open.

References

1. Afraimovich, V.S., Bykov, V.V., Shil'nikov, L.P.: On structurally unstable attracting limit sets of Lorenz attractor type. Trudy Moskov. Mat. Obshch. **44**, 150–212 (1982); English translation in Trans. Moscow Math. Soc. **2**, 153–216 (1983)
2. Barge, M.: Horseshoe maps and inverse limits. Pacific J. Math. **121**(1), 29–39 (1986)
3. Guckenheimer, J., Williams, R.F.: Structural stability of Lorenz attractors. Inst. Hautes Études Sci. Publ. Math. **50**, 59–72 (1979)
4. Klinshpont, N.E.: A topological invariant of the Lorenz attractor. Uspekhi Mat. Nauk **47**(2), 195–196 (1992); English translation in Russian Math. Surveys **47**(2), 221–223 (1992)
5. Klinshpont, N.E.: The topological invariant of the Lorenz's type attractors. Appendix to Plykin, R.V., Zhirov, A.J.: Some problems of attractors of dynamical systems. Topol. Appl. **54**(1–3), 39–46 (1993)
6. Klinshpont, N.E.: On the topological classification of Lorenz-type attractors. Matematicheskii Sbornik. **197**(4), 75–122 (2006); English translation in Sbornik Math. **197**(4), 547–593 (2006)
7. Lorenz, E.N. :Deterministic nonperiodic flow. J. Atmos. Sci. **20**(2), 130–141 (1963)
8. McLaughlin, J.B., Martin, P.C.: Transition to turbulence of statically stressed fluids. Phys. Rev. Lett. **39**(20), 1189–1192 (1974)
9. Plykin, R.V.: To a problem of topological classification of strange attractors of dynamic systems. Uspekhi Mat. Nauk. **57**(6), 123–166 (2002). English translation: Russian Math. Surveys **57**(6), 1163–1205 (2002)
10. Plykin, R.V., Zhirov, A.J.: Some problems of attractors of dynamical systems. Topol. Appl. **54**(1–3), 39–46 (1993)
11. Rand, D.: The topological classification of Lorenz attractors. Math. Proc. Camb. Phil. Soc. **83**(3), 451–460 (1978)
12. Tucker, W.: A rigorous ODE solver and Smale's 14th problem. Found. Comput. Math. **2**(1), 53–117 (2002)
13. Viana, M.: What's new on Lorenz strange attractors? Math. Intell. **22**(3), 6–19 (2000)
14. Watkins, W.T.: Homeomorphic classification of certain inverse limit spaces with open bonding maps. Pacific J. Math. **103**(2), 589–601 (1982).
15. Williams, R.F.: The structure of Lorenz attractors. In: Turbulence Seminar (Univ. Calif., Berkeley, CA, 1976/77). Lecture Notes in Mathematics, vol. 615, pp. 94–112. Springer, Berlin (1977)
16. Williams, R.F.: The structure of Lorenz attractors. Inst. Hautes Études Sci. Publ. Math. **50**, 73–99 (1979)
17. Yorke, J.A., Yorke, E.D.: Metastable chaos: The transition to sustained chaotic behavior in the Lorenz model. J. Statist. Phys. **21**(3), 263–377 (1979)

Chapter 13
A Nonlinear Map for the Decay to Equilibrium of Ideal Gases

Ricardo López-Ruiz and Elyas Shivanian

Abstract In this communication, we present an operator that governs the discrete time evolution of the velocity distribution of an out-of-equilibrium ideal gas. This nonlinear map, which conserves the momentum and the energy of the ideal gas, has the Maxwellian velocity distribution (MVD) as an asymptotic equilibrium. All these properties are presented in this communication.

13.1 Introduction

A framework inspired in economic systems that has recently been proposed [3] to explain the attractivity, and then the ubiquity, of the exponential (Boltzmann–Gibbs) distribution (BGD) reads as follows: Let $p(m)\mathrm{d}m$ denote the PDF (*probability density function*) of money in a multiagent economic system, i.e., the probability of finding an agent of the ensemble with money between m and $m+\mathrm{d}m$. Consider now the discrete-time evolution of an initial money distribution $p_0(m)$ at each time step n under the action of an operator $\mathscr{T}$, which represents the average effect on the system of many random binary interactions (with number of the order of the system size) between pairs of agents exchanging their money. Thus, the system evolves

R. López-Ruiz (✉)
DIIS-BIFI, University of Zaragoza, E-50009 Zaragoza, Spain
e-mail: rilopez@unizar.es

E. Shivanian
Department of Mathematics, Imam Khomeini International University, Qazvin, 34149-16818, Iran
e-mail: shivanian@ikiu.ac.ir

C. Grácio et al. (eds.), *Nonlinear Maps and their Applications*, Springer Proceedings in Mathematics & Statistics 57, DOI 10.1007/978-1-4614-9161-3_13,

from time n to time $n+1$ to asymptotically reach the equilibrium wealth distribution $p_f(m)$, i.e.,

$$\lim_{n\to\infty} \mathscr{T}^n\left(p_0(m)\right) \to p_f(x) = BGD = \delta e^{-\delta x} \qquad with \qquad \delta =< p_0 >^{-1}.$$

In this case, $p_f(m)$ is the exponential distribution (BGD) with the same average wealth $< p_f >= \delta^{-1}$ as the initial one $< p_0 >$, due to the local and total money conservation [6]. The mathematical properties of the operator $\mathscr{T}$ have been presented in [2]. Hence, this framework provides evidence that the BGD is the equilibrium distribution in an economic gaslike system with symmetric exchange rules. Also, it is found in this case that the BGD is asymptotically reached independently of the initial wealth distribution given to the system, a point of view that to date was possibly lacking in the literature.

In this work, we extend this perspective to another problem of the same statistical nature. Our goal is to explain the ubiquity of the Maxwellian velocity distribution (MVD) in ideal gases [1,4]. In the next section, we show how to obtain an operator T in the space of velocity distributions in order to explain the decay of any initial velocity distribution to the MVD. Then the dynamical properties of this operator T will be sketched and some examples showing its dissipative behavior will be given. Finally, we present our conclusions.

13.2 The Nonlinear Map T

Consider an ideal gas with particles of unit mass in three-dimensional (3D) space. As long as there is not a privileged direction in the equilibrium, we can take any direction in the space and study the discrete-time evolution of the velocity distribution in that direction. Let us call this direction U. We can complete a Cartesian system with two additional orthogonal directions V, W. If $p_n(u)du$ represents the probability of finding a particle of the gas with velocity component in the direction U located between u and $u+du$ at time n, then the probability of having at this time n a particle with a 3D velocity (u,v,w) will be $p_n(u)p_n(v)p_n(w)$.

The particles of the gas collide, and after a number of interactions of order the system size, a new velocity distribution is attained at time $n+1$. Concerning the interaction of particles with the bulk of the gas, we make two simplistic and realistic assumptions in order to obtain the probability of having a velocity x in the direction U at time $n+1$: (1) Only those particles with an energy greater than x^2 at time n can contribute to this velocity x in the direction U, that is, all those particles whose velocities (u,v,w) satisfy $u^2+v^2+w^2 \geq x^2$. (2) The new velocities after collision are equally distributed in their permitted ranges, that is, particles with velocity (u,v,w) can generate maximal velocities $\pm U_{\max} = \pm\sqrt{u^2+v^2+w^2}$. Then the allowed range of velocities $[-U_{\max}, U_{\max}]$ is of size $2|U_{\max}|$, and the contributing probability of these particles to the velocity x will be $p_n(u)p_n(v)p_n(w)/(2|U_{\max}|)$. Taking all this together, we finally get the expression for the evolution operator T:

$$p_{n+1}(x) = Tp_n(x) = \int\int\int_{u^2+v^2+w^2 \geq x^2} \frac{p_n(u)p_n(v)p_n(w)}{2\sqrt{u^2+v^2+w^2}}\,\mathrm{d}u\mathrm{d}v\mathrm{d}w\,.$$

Let us note that we have not made any supposition about the type of interactions or collisions between the particles and that to some extent, the equivalent of the Boltzmann hypothesis of *molecular chaos* would be the two simplistic assumptions we have stated on the interaction of particles with the bulk of the gas. But now a clearer and more understandable framework than those usually presented in the literature [4] appears on the scene. In fact, we are going to show that the operator T conserves in time the energy and the null momentum of the gas. Moreover, for any initial velocity distribution, the system tends toward its equilibrium, i.e., toward the MVD. This means that

$$\lim_{n\to\infty} T^n\left(p_0(x)\right) \to p_f(x) = MVD \quad \text{(1D case)}.$$

13.3 Some Properties of the Operator T

In order to prove some of the properties of the operator T, we provide an adequate mathematical framework. First, we introduce the space L_1^+ of positive functions (one-dimensional velocity distributions) on the real axis,

$$L_1^+(\Re) = \{p(x) : \Re \to \Re^+ \cup \{0\},\ \ ||p|| < +\infty\},$$

with norm $||\cdot||$ defined by

$$||p|| = \int_{-\infty}^{+\infty} p(x)dx.$$

Therefore, for each $p \in L_1^+(\Re)$, we have $||p|| \geq 0$.

Consider now the set $B \subset L_1^+(\Re)$ of PDFs (probability density functions), i.e., $B = \{p \in L_1^+(\Re) : ||p|| = 1\}$. Then if $p \in B$, we have $Tp \in B$, i.e., the action of T on B conserves the number of particles in the gas. This is a consequence of the next theorem.

Theorem 1. *For all* $p \in L_1^+(\Re)$, *we have* $||Tp|| = ||p||^3$.

Proof. Suppose that $p \in L_1^+(\Re)$. Then

$$||Tp|| = \int_{-\infty}^{+\infty} Tp(x)dx =$$

$$= \int_{-\infty}^{+\infty} \left(\int\int\int_{u^2+v^2+w^2 \geq x^2} \frac{p(u)p(v)p(w)}{2\sqrt{u^2+v^2+w^2}} \mathrm{d}u\mathrm{d}v\mathrm{d}w \right) \mathrm{d}x =$$

$$= \int_{-\infty}^{+\infty} \int_{-\infty}^{+\infty} \int_{-\infty}^{+\infty} \left(\int_{-\sqrt{u^2+v^2+w^2}}^{+\sqrt{u^2+v^2+w^2}} \frac{p(u)p(v)p(w)}{2\sqrt{u^2+v^2+w^2}} \mathrm{d}x \right) \mathrm{d}u\mathrm{d}v\mathrm{d}w =$$

$$= \int_{-\infty}^{+\infty} \int_{-\infty}^{+\infty} \int_{-\infty}^{+\infty} p(u)p(v)p(w)\mathrm{d}u\mathrm{d}v\mathrm{d}w =$$

$$= \int_{-\infty}^{+\infty} p(u)\mathrm{d}u \int_{-\infty}^{+\infty} p(v)\mathrm{d}v \int_{-\infty}^{+\infty} p(w)\mathrm{d}w = \|p\|^3 .$$

□

The zero total momentum of the gas is conserved in its time evolution under the action of T, as proved in the next theorem.

Theorem 2. *The mean value of the velocity in the recurrence $p_n = T^n p_0$ is conserved in time. In fact, it is null for all n:*

$$< x, Tp > = < x, T^2 p > = < x, T^3 p > = \cdots = < x, T^n p > = \cdots = 0,$$

where

$$< f, g > = \int_{-\infty}^{+\infty} f(x) g(x) \mathrm{d}x .$$

Proof. Assume that $p \in B$. Then

$$< x, Tp > = \int_{-\infty}^{+\infty} xTp(x)dx =$$

$$= \int_{-\infty}^{+\infty} x \left(\int\int\int_{u^2+v^2+w^2 \geq x^2} \frac{p(u)p(v)p(w)}{2\sqrt{u^2+v^2+w^2}} \mathrm{d}u\mathrm{d}v\mathrm{d}w \right) \mathrm{d}x =$$

$$= \int_{-\infty}^{+\infty} \int_{-\infty}^{+\infty} \int_{-\infty}^{+\infty} \frac{p(u)p(v)p(w)}{2\sqrt{u^2+v^2+w^2}} \left(\int_{-\sqrt{u^2+v^2+w^2}}^{+\sqrt{u^2+v^2+w^2}} x\mathrm{d}x \right) \mathrm{d}u\mathrm{d}v\mathrm{d}w =$$

$$= \int_{-\infty}^{+\infty} \int_{-\infty}^{+\infty} \int_{-\infty}^{+\infty} \frac{p(u)p(v)p(w)}{2\sqrt{u^2+v^2+w^2}} \left[\frac{x^2}{2} \right]_{-\sqrt{u^2+v^2+w^2}}^{+\sqrt{u^2+v^2+w^2}} \mathrm{d}u\mathrm{d}v\mathrm{d}w = 0 .$$

We know that if $p \in B$ then $Tp \in B$, and therefore $< x, T^n p > = 0 \ \forall n$. □

Also, the total energy of the gas is conserved in time as shown in the following theorem.

Theorem 3. *For every $p \in B$, we have*

$$< x^2, p > = < x^2, Tp > = < x^2, T^2 p > = < x^2, T^3 p > = \cdots = < x^2, T^n p > = \cdots .$$

Proof. Assume that $p \in B$. Then

$$< x^2, Tp > = \int_{-\infty}^{+\infty} x^2 Tp(x)dx =$$

$$= \int_{-\infty}^{+\infty} x^2 \left(\iiint_{u^2+v^2+w^2 \geq x^2} \frac{p(u)p(v)p(w)}{2\sqrt{u^2+v^2+w^2}} \mathrm{d}u\mathrm{d}v\mathrm{d}w \right) \mathrm{d}x =$$

$$= \int_{-\infty}^{+\infty} \int_{-\infty}^{+\infty} \int_{-\infty}^{+\infty} \frac{p(u)p(v)p(w)}{2\sqrt{u^2+v^2+w^2}} \left(\int_{-\sqrt{u^2+v^2+w^2}}^{+\sqrt{u^2+v^2+w^2}} x^2 \,\mathrm{d}x \right) \mathrm{d}u\mathrm{d}v\mathrm{d}w =$$

$$= \int_{-\infty}^{+\infty} \int_{-\infty}^{+\infty} \int_{-\infty}^{+\infty} \frac{p(u)p(v)p(w)}{2\sqrt{u^2+v^2+w^2}} \left[\frac{x^3}{3} \right]_{-\sqrt{u^2+v^2+w^2}}^{+\sqrt{u^2+v^2+w^2}} \mathrm{d}u\mathrm{d}v\mathrm{d}w.$$

$$= \int_{-\infty}^{+\infty} \int_{-\infty}^{+\infty} \int_{-\infty}^{+\infty} \frac{p(u)p(v)p(w)}{2\sqrt{u^2+v^2+w^2}} \frac{2}{3} (u^2+v^2+w^2)^{\frac{2}{3}} \mathrm{d}u\mathrm{d}v\mathrm{d}w =$$

$$= \int_{-\infty}^{+\infty} \int_{-\infty}^{+\infty} \int_{-\infty}^{+\infty} \frac{1}{3} (u^2+v^2+w^2) \, p(u)p(v)p(w) \mathrm{d}u\mathrm{d}v\mathrm{d}w =$$

$$= \frac{1}{3} \int_{-\infty}^{+\infty} u^2 p(u)\mathrm{d}u + \frac{1}{3} \int_{-\infty}^{+\infty} v^2 p(v)\mathrm{d}v + \frac{1}{3} \int_{-\infty}^{+\infty} w^2 p(w)\mathrm{d}w =$$

$$= \int_{-\infty}^{+\infty} x^2 p(x)\mathrm{d}x = < x^2, p > .$$

We know that if $p \in B$ then $Tp \in B$, and therefore $< x^2, T^n p > = < x^2, p > \forall n$. □

The Maxwellian (Gaussian) distribution is the fixed point of the operator T, as we prove here.

Theorem 4. *The one-parameter family of PDFs* $p_\alpha(x) = \sqrt{\frac{\alpha}{\pi}} e^{-\alpha x^2}$, $\alpha \geq 0$ *consists of fixed points of the operator* T*. In other words,* $Tp_\alpha = p_\alpha$.

Proof. When $\alpha = 0$, then $p_\alpha = 0$, which is clearly a fixed point of T. Suppose now that $\alpha \neq 0$,

$$Tp_\alpha = \iiint_{u^2+v^2+w^2 \geq x^2} \frac{p_\alpha(u)p_\alpha(v)p_\alpha(w)}{2\sqrt{u^2+v^2+w^2}} \mathrm{d}u\mathrm{d}v\mathrm{d}w =$$

$$= \iiint_{u^2+v^2+w^2 \geq x^2} \frac{\frac{\alpha}{\pi}\sqrt{\frac{\alpha}{\pi}} e^{-\alpha(u^2+v^2+w^2)}}{2\sqrt{u^2+v^2+w^2}} \mathrm{d}u\mathrm{d}v\mathrm{d}w.$$

On transforming the integral region to spherical coordinates by the change of variables

$$u = r \sin\psi \cos\theta, v = r \sin\psi \sin\theta, w = r\cos\psi,$$

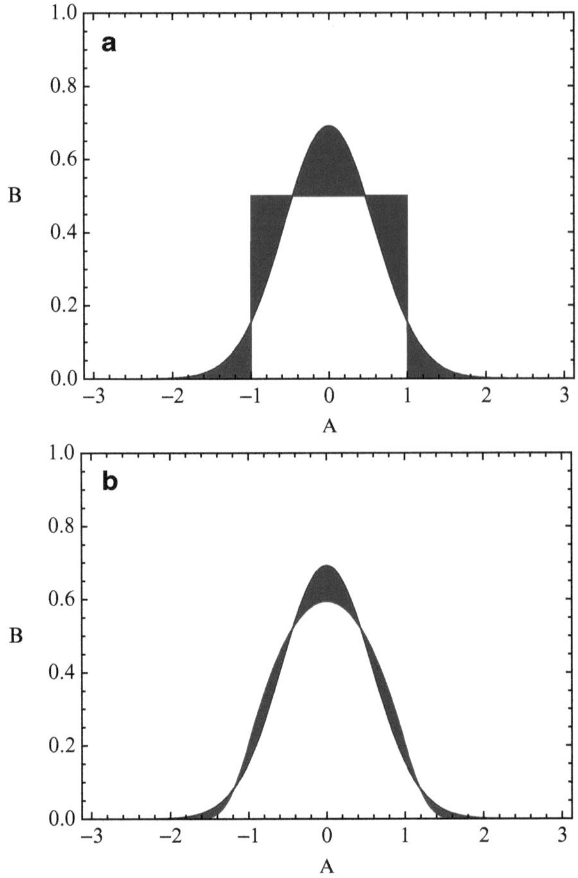

Fig. 13.1 (**a**) $p(x)$ and $\mu(x)$, (**b**) $Tp(x)$ and $\mu(x)$

the proof becomes straightforward:

$$Tp_\alpha = \int_{|x|}^{+\infty}\int_0^{\pi}\int_0^{2\pi}\frac{\alpha}{2\pi}\sqrt{\frac{\alpha}{\pi}}e^{-\alpha r^2} r\sin\psi \mathrm{d}\theta \mathrm{d}\psi \mathrm{d}r =$$

$$= \int_{|x|}^{+\infty}\int_0^{\pi}\alpha\sqrt{\frac{\alpha}{\pi}}e^{-\alpha r^2} r\sin\psi \mathrm{d}\psi \mathrm{d}r = \int_{|x|}^{+\infty}2\alpha\sqrt{\frac{\alpha}{\pi}}re^{-\alpha r^2}\mathrm{d}r =$$

$$= \sqrt{\frac{\alpha}{\pi}}\left[-e^{-\alpha r^2}\right]_{|x|}^{+\infty} = \sqrt{\frac{\alpha}{\pi}}e^{-\alpha x^2} = p_\alpha \,.$$

□

Other additional properties, as for instance the increase of entropy with time, can also be determined for the operator T. To end this communication, let us recall at this point the main result that resumes the behavior of the system [5]:

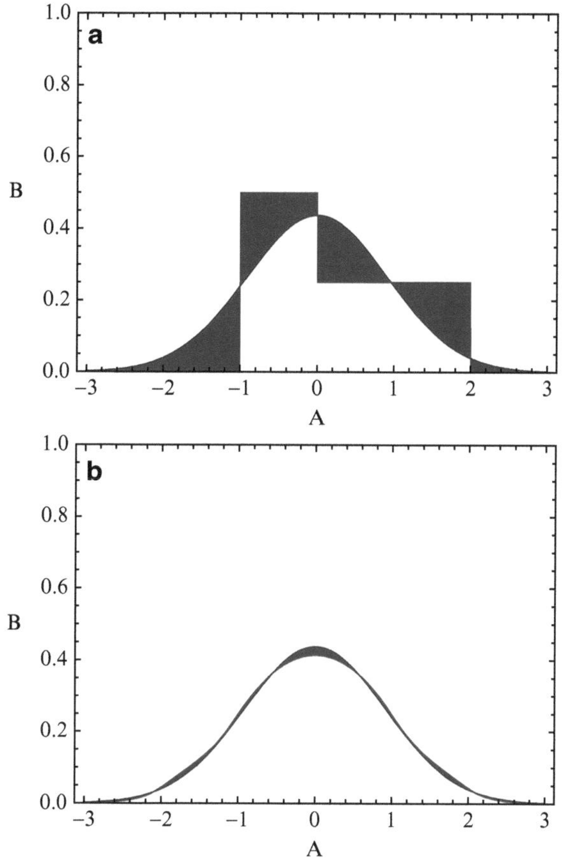

Fig. 13.2 (**a**) $p(x)$ and $\mu(x)$, (**b**) $Tp(x)$ and $\mu(x)$

Conjecture: For every $p \in B$, with B the set of functions whose norm $\|p\|$ is equal to unity, with finite mean energy, $< x^2, p >$, and satisfying $\lim_{n\to\infty} \|T^n p(x) - \mu(x)\| = 0$, the limit $\mu(x)$ is the fixed point $p_\alpha(x) = \sqrt{\frac{\alpha}{\pi}}\, e^{-\alpha x^2}$, with $\alpha = (2 < x^2, p >)^{-1}$.

In physical terms, this means that every initial velocity distribution of the gas decays to the Maxwellian distribution, which is just the fixed point of the dynamics. Recalling that $< x^2, p >= k\tau$, with k the Boltzmann constant and τ the temperature of the gas, and introducing the mass m of the particles, let us observe that the MVD is recovered in its 3D format:

$$MVD = p_\alpha(u)p_\alpha(v)p_\alpha(w) = \left(\frac{m\alpha}{\pi}\right)^{\frac{3}{2}} e^{-m\alpha(u^2+v^2+w^2)} \quad with \quad \alpha = (2k\tau)^{-1}.$$

Some examples showing this behavior are depicted in Figs. 13.1 and 13.2. In these figures, the horizontal axis represents the x-coordinate of the distributions, and on the vertical axis, the distributions $p(x)$, $Tp(x)$, and $\mu(x)$ are plotted. Observe that the convergence of the T-iterations of any distribution $p(x)$ to its Gaussian limit $\mu(x)$ is very fast.

13.4 Conclusions

In this paper, we have presented a nonlinear map acting on the velocity distribution space of ideal gases that gives an account of the decay of an out-of-equilibrium velocity distribution toward the Maxwellian distribution. Some properties and figures concerning its dynamical behavior have also been shown.

References

1. Boltzmann, L.: Lectures on Gas Theory (translated by S.G. Brush). Dover Publications, New York (1995)
2. Lopez, J.L., Lopez-Ruiz, R., Calbet, X.: Exponential wealth distribution in a random market: A rigorous explanation. J. Math. Anal. Appl. **386**, 195–204 (2012); also at arXiv:1103.5703 (2011)
3. Lopez-Ruiz, R., Lopez, J.L., Calbet, X.: Exponential wealth distribution: A new approach from functional iteration theory. In: ESAIM Proceedings, ECIT-2010 Conference, September (2010); also at arXiv:1103.1501 (2011)
4. Maxwell, J.C.: The Scientific Papers of James Clerk Maxwell (ed. by Niven, W.D.). Paper Number XX, vol. I, II. Dover Publications, New York (2003)
5. Shivanian, E., Lopez-Ruiz, R.: A new model for ideal gases: Decay to the Maxwellian distribution. Physica A **391**, 2600–2607 (2012); also at arXiv:1105.4813 (2011)
6. Yakovenko, V.M.: Econophysics, statistical mechanics approach to. In: Meyers, R.A. (ed.) Encyclopedia of Complexity and System Science. Springer, Berlin (2009)

Chapter 14
On the Dynamics of a Cubic p-adic Polynomial

Sérgio Mendes

Abstract We investigate the dynamics of a cubic polynomial in the p-adic setting. Specifically, we describe the Siegel disks and the basin of attraction for the dynamical system using a result of Albeverio et al.

14.1 Introduction

The study of dynamical systems in the p-adic setting has recently become very active. Research in this area includes both topological and measurable dynamics. Due to the nature of the nonarchimedean world, arithmetic and number-theoretic problems have been addressed by several authors in connection with dynamical systems; see Silverman's book [9] and the bibliography therein.

The purpose of considering p-adic dynamical systems is also motivated by a new trend in applied mathematics. In recent years, natural sciences such as physics and biology have been using not only the classical archimedean fields $\mathbb{R}$ and $\mathbb{C}$ but also the nonarchimedean p-adic field $\mathbb{Q}_p$ for their models. See, for instance, [4, 10].

Several articles have dealt with the study of concrete examples, in which the map is typically either a polynomial or a rational function. Quite specifically, these examples include the study of monomial dynamical systems $f(x) = x^n$ [3] and perturbed monomial dynamical systems $f_q(x) = x^n + q(x)$ (here $q(x)$ is a polynomial whose coefficients have small p-adic absolute value) [7], fractional dynamical systems of the form $f(x) = \frac{x+a}{bx+c} (a, b, c \in \mathbb{C}_p)$ [6] (where $\mathbb{C}_p$ denotes the complex p-adic numbers), and fractional dynamical systems of the form $f(x) = \frac{ax^2}{bx+1} (a, b \in \mathbb{C}_p)$. Quadratic dynamical systems $f(x) = x^2 + c, c \in \mathbb{Q}_p$ were considered, among others, by Shabat in [8] and by Dremov et al. [2]. In [5], Mukhamedov and Mendes

---

S. Mendes (✉)
ISCTE – Lisbon University Institute, Av. das Forças Armadas, 1649-026 Lisbon, Portugal
e-mail: sergio.mendes@iscte.pt

C. Grácio et al. (eds.), *Nonlinear Maps and their Applications*, Springer Proceedings in Mathematics & Statistics 57, DOI 10.1007/978-1-4614-9161-3_14,

considered a cubic dynamical system of the form $f(x) = x^3 + ax^2$, $a \in \mathbb{Q}_p \backslash \mathbb{Z}_p^{\times}$ (here $\mathbb{Z}_p$ denotes the ring of p-adic integers and $\mathbb{Z}_p^{\times}$ its group of units). This list is obviously far from exhaustive.

A problem addressed in the work of Mukhamedov and Mendes [5] is the fact that fixed points are given by square roots, and, as is well known, square roots do not always exist in $\mathbb{Q}_p$. Nevertheless, the authors gave sufficient conditions for the existence of square roots in their example.

In this note, we shall study the dynamical system associated to the cubic polynomial $f(x) = x^3 + ax^2 + x$ ($a \in \mathbb{Q}_p$). This example is quite similar to that of Mukhamedov and Mendes. Essentially, it differs in the term of degree one and in the fact that we do not need to impose restrictions on the parameter $a \in \mathbb{Q}_p$, i.e., we allow the possibility of a being a unit in the ring of p-adic integers, which means simply we that allow the possibility of $|a|_p = 1$. This is due to the fact that the fixed points of $f(x) = x^3 + ax^2 + x$ are no longer square roots, and therefore they always exist in $\mathbb{Q}_p$.

This article is organized as follows. In Sect. 14.2, we recall some basic definitions about p-adic numbers and dynamical systems. In Sect. 14.3, we recall a fundamental result of Albeverio et al. [1]. The rest of Sect. 14.3 is devoted to the cubic polynomial $f(x) = x^3 + ax^2 + x$. Section 14.4 is dedicated to the study of indifferent points and their Siegel disks; see Theorems 2 and 3. We also study an attractor that occurs only for $p = 2$ and $|a|_2 = 1$, and we describe its basin of attraction; see Theorem 4.

14.2 Background

Let $\mathbb{Q}_p$ be the field of p-adic numbers. Denote the greatest common divisor of integers a and b by (a,b). Every rational number $x \neq 0$ can be written uniquely in the form $x = p^r \frac{m}{n}$, where $r, m \in \mathbb{Z}$, $n \in \mathbb{N}$, p is a fixed prime, and $(m,n) = (m,p) = (n,p) = 1$. The p-adic norm of x is given by

$$|x|_p = \begin{cases} p^{-r}, & x \neq 0, \\ 0, & x = 0. \end{cases}$$

Moreover, $|\cdot|_p$ satisfies the usual properties of a norm:

(i) $|x|_p \geq 0$ and $|x|_p = 0$ if and only if $x = 0$;
(ii) $|xy|_p = |x|_p |y|_p$;

and the *strong triangle inequality*

(iii) $|x+y|_p \leq \max\{|x|_p, |y|_p\}$.

From (iii), we can deduce the following:

$$\text{If } |x|_p \neq |y|_p, \text{ then } |x-y|_p = \max\{|x|_p, |y|_p\}. \tag{14.1}$$

$$\text{If } |x|_p = |y|_p, \text{ then } |x-y|_p \le |2x|_p. \tag{14.2}$$

The completion of $\mathbb{Q}$ with respect to the p-adic norm gives rise to the field of p-adic numbers $\mathbb{Q}_p$.

We shall use the following notation:

$$B_r(a) = \{x \in \mathbb{Q}_p : |x-a|_p < r\},\ \overline{B}_r(a) = \{x \in \mathbb{Q}_p : |x-a|_p \le r\},$$

$$S_r(a) = \{x \in \mathbb{Q}_p : |x-a|_p = r\}.$$

Remark 1. The ball $\overline{B}_1(0)$ is called the *ring of p-adic integers* and is usually denoted by $\mathbb{Z}_p$. The sphere $S_1(0)$ is the group of units of the ring $\mathbb{Z}_p$ and is denoted by $\mathbb{Z}_p^{\times}$.

A function $f : B_r(a) \to \mathbb{Q}_p$ is said to be *analytic* if it can be represented by a power series

$$f(x) = \sum_{n\ge 0}^{\infty} a_n (x-a)^n, a_n \in \mathbb{Q}_p,$$

that converges uniformly in $B_r(a)$.

We recall now some standard notions of dynamical systems. Set $U = B_r(a) \subset \mathbb{Q}_p$ for a fixed $r > 0$ and $a \in \mathbb{Q}_p$. Let $f : U \longrightarrow U$ be an analytic function. Given $x_0 \in U$, we set $x^{(n)} = f^n(x_0)$, where $f^n = f \circ f \circ \ldots \circ f$ denotes the nth iterate of f.

If $f(x_0) = x_0$, then x_0 is called a *fixed point* of the dynamical system (f, U).

Definition 1. A fixed point x_0 of (f, U) is called an attractor if there exists a neighborhood U_{x_0} of x_0 such that for every $y \in U_{x_0}$, $\lim_{n\to\infty} y^{(n)} = x_0$.

If x_0 is an attractor, its *basin of attraction* is the set

$$\mathscr{A}(x_0) = \{y \in \mathbb{Q}_p : y^{(n)} \to x_0, n \to \infty\}.$$

Definition 2. A fixed point x_0 of (f, U) is called a repeller if there exists a neighborhood U_{x_0} of x_0 such that $|f(x) - x_0|_p > |x - x_0|_p$ for every $x \in U_{x_0}$, $x \ne x_0$.

The ball $B_r(x_0) \subset U$ is said to be a *Siegel disk* if for each $\rho < r$,

$$x \in S_\rho(x_0) \Rightarrow x^{(n)} \in S_\rho(x_0), \quad n = 1, 2, \ldots.$$

The union of all Siegel disks with center at x_0 is called the *maximum Siegel disk* and is denoted by $\mathscr{S}\mathscr{I}(x_0)$.

A peculiarity of nonarchimedean topology is that every point in a disk is a center of the disk. As a consequence, different fixed points may have the same Siegel disk.

A simple criterion for classifying fixed points is given by the first derivative as follows. Let x_0 be a fixed point of the dynamical system (f, U) and write $\lambda = f'(x_0)$. Then x_0 is called *attractive* if $0 \le |\lambda|_p < 1$, *indifferent* if $|\lambda|_p = 1$, and *repelling* if $|\lambda|_p > 1$.

14.3 The Dynamical System $f(x) = x^3 + ax^2 + x$, $a \in \mathbb{Q}_p$

In this section, we concentrate on the dynamical system $f : \mathbb{Q}_p \to \mathbb{Q}_p$, $f(x) = x^3 + ax^2 + x$, with $a \in \mathbb{Q}_p$. Solving $f(x) = x$, we conclude that there are two fixed points, $x_0 = 0$ and $x_1 = -a$. Using the derivative, we may now classify the fixed points x_0 and x_1.

Since $f'(x_0)_p = 1$, we conclude that $|f'(x_0)|_p = 1$, and so $x_0 = 0$ is indifferent for every $a \in \mathbb{Q}_p$.

On the other hand, $|f'(x_1)|_p = |a^2+1|_p$. It follows that $|f'(x_1)|_p = 1$ if $|a|_p < 1$, $|f'(x_1)|_p > 1$ if $|a|_p > 1$, and $|f'(x_1)|_2 = \frac{1}{2} < 1$ if $|a|_2 = 1$. Therefore, we conclude that $x_1 = -a$ is indifferent for $|a|_p < 1$, a repeller for $|a|_p > 1$, and attractive for $|a|_2 = 1$.

Our main purpose is to study in detail each of the above cases. We need the following result from [1].

Theorem 1 ([1]). *Let x_0 be a fixed point of an analytic function $f : U \to U$. The following assertions hold:*

(i) If x_0 is an attractive point of f, then it is an attractor of the dynamical system (f, U). If $r > 0$ satisfies the inequality

$$\max_{n\geq 1} \left| \frac{1}{n!} \frac{d^n f}{dx^n}(x_0) \right|_p r^{n-1} < 1 \tag{14.3}$$

and $B_r(x_0) \subset U$, then $B_r(x_0) \subset \mathscr{A}(x_0)$.

(ii) If x_0 is an indifferent point of f, then it is the center of a Siegel disk. If $r > 0$ satisfies the inequality

$$max_{n\geq 2} \left| \frac{1}{n!} \frac{d^n f}{dx^n}(x_0) \right|_p r^{n-1} < 1 \tag{14.4}$$

and $B_r(x_0) \subset U$, then $B_r(x_0) \subset \mathscr{S}\mathscr{I}(x_0)$.

(iii) If x_0 is a repelling point of f, then x_0 is a repeller of the dynamical system (f, U).

Now let x_i be a fixed point of $f(x) = x^3 + ax^2 + x$, $i = 0, 1$. We may represent f as follows:

$$f(x) = x_i + f'(x_i)(x - x_i) + \frac{f''(x_i)}{2!}(x - x_i)^2 + \frac{f'''(x_i)}{3!}(x - x_i)^3.$$

The following result is identical to [5, Lemma 4.1].

Lemma 1. *Let x_i be a fixed point of $f(x) = x^3 + ax^2 + x$, $i = 0, 1$. If*

$$\max\{|3x_i + a|_p |x - x_i|_p, |x - x_i|_p^2\} < |f'(x_i)|_p,$$

then

$$|f(x)-x_i|_p = |f'(x_i)|_p |x-x_i|_p.$$

Proof. Computing $f''(x_i)$ and $f'''(x_i)$, we have

$$|f(x)-x_i|_p = |f'(x_i)(x-x_i)+(3x_i+a)(x-x_i)^2+(x-x_i)^3|_p$$

$$= |x-x_i|_p |f'(x_i)+(3x_i+a)(x-x_i)+(x-x_i)^2|_p,$$

and the result follows from the strong triangle inequality. □

Combining Theorem 1 and Lemma 1, we obtain the following result.

Corollary 1. *Let x_i be a fixed point of the function $f(x)=x^3+ax^2+x$, $i=0,1$. The following assertions hold:*

(i) If x_i is an attractive point of f, then it is an attractor of the dynamical system. If $r>0$ satisfies the inequality

$$\max\{|3x_i+a|_p r, r^2\} < 1, \tag{14.5}$$

then $B_r(x_i) \subset \mathscr{A}(x_i)$.

(ii) If x_i is an indifferent point of f, then it is the center of a Siegel disk. If $r>0$ satisfies the inequality (14.5), then $B_r(x_i) \subset \mathscr{S}\mathscr{I}(x_i)$.

(iii) If x_i is a repelling point of f, then x_i is a repeller of the dynamical system.

14.4 Siegel Disks and Attractors

This section is devoted to describing the Siegel disks for $f(x)=x^3+ax^2+x$, $a \in \mathbb{Q}_p$. We start with the fixed point $x_0=0$. As we have seen before, $|f'(x_0)|_p = 1$, for every $a \in \mathbb{Q}_p$. Therefore, we have the following.

Theorem 2. *Let $x_0=0$. Then, x_0 is indifferent for every $a \in \mathbb{Q}_p$. Moreover:*

(i) If $|a|_p<1$ and $p \neq 2$, then $\mathscr{S}\mathscr{I}(x_0)=\overline{B}_1(x_0)$. If $p=2$ and $|a|_2<1$, then $\mathscr{S}\mathscr{I}(x_0)=B_1(x_0)$.

(ii) If $|a|_p=1$ for every prime p, then $\mathscr{S}\mathscr{I}(x_0)=B_1(x_0)$.

(iii) If $|a|_p>1$, then $\mathscr{S}\mathscr{I}(x_0)=B_{\frac{1}{|a|_p}}(x_0)$.

Proof. (i) Suppose $|a|_p<1$. Condition (14.5) becomes simply

$$\max\{|a|_p r, r^2\} < 1$$

and is satisfied if $r<1$. From Corollary 1, we have $B_1(x_0) \subset \mathscr{S}\mathscr{I}(x_0)$.

Suppose $|x-x_0|_p = |x|_p = 1$. Then

$$|f(x)-x_0|_p = |f(x)|_p = |x|_p |x^2+ax+1|_p = 1,$$

provided that $p \neq 2$, and so $f(S_1(x_0)) \subset S_1(x_0)$. We conclude that the maximum Siegel disk in this case is $\mathcal{SI}(x_0) = \overline{B}_1(x_0)$.

Now if $p = 2$ and $|x|_2 = 1$, we have

$$|f(x) - x_0|_2 = |f(x)|_2 \leq \frac{1}{2}.$$

Therefore, $\mathcal{SI}(x_0) = B_1(x_0)$ for $p = 2$.

(ii) Suppose $|a|_p = 1$. Condition (14.5) becomes

$$\max\{r, r^2\} < 1$$

and is satisfied if $r < 1$. Hence, Corollary 1 implies $B_1(x_0) \subset \mathcal{SI}(x_0)$. It suffices to prove that for every $\varepsilon < 1$, the equality

$$f(S_\varepsilon(x_0)) = S_\varepsilon(x_0)$$

holds. If $|x - x_0|_p = |x|_p = \varepsilon$, then

$$|f(x) - x_0|_p = |f(x)|_p = |x^3 + ax^2 + x|_p = |x|_p |x^2 + ax + 1|_p$$
$$= \max\{|x|_p |x + a|_p, 1\} = \max\{\varepsilon, 1\} = 1,$$

and the result follows.

(iii) Consider now $|a|_p > 1$. Condition (14.5) is satisfied if $r < \frac{1}{|a|_p}$. By Corollary 1, for every $r < \frac{1}{|a|_p}$, $B_r(x_0) \subset \mathcal{SI}(x_0)$.

Take $-\frac{1}{a} \in \overline{B}_{\frac{1}{|a|_p}}(x_0)$. Then $f(-\frac{1}{a}) = -\frac{1}{a^3}$ and

$$|f(-\frac{1}{a})|_p = \frac{1}{|a|_p^3} \neq \frac{1}{|a|_p}.$$

Therefore, $f(S_{\frac{1}{|a|_p}}(x_0)) \not\subseteq S_{\frac{1}{|a|_p}}(x_0)$, and so $\mathcal{SI}(x_0) = B_{\frac{1}{|a|_p}}(x_0)$. □

Now we concentrate on the fixed point $x_1 = -a$. As we have seen before, $|f'(x_1)|_p = 1$ whenever $|a|_p < 1$. We therefore have the following theorem.

Theorem 3. *Let $x_1 = -a$. If $|a|_p < 1$, then x_1 is indifferent for $|a|_p < 1$, and we have*

$$\mathcal{SI}(x_1) = B_1(x_1).$$

Proof. Consider $|a|_p < 1$. Condition (14.5) becomes

$$\max\{|2a|_p r, r^2\} < 1$$

and is satisfied if $r < 1$. Hence from Corollary 1, $B_r(x_0) \subset \mathcal{SI}(x_0)$, for every $r < 1$.

Suppose $|x + a|_p = 1$. Then $|f(-a) + a|_p = 0$, and so $\mathcal{SI}(x_1) = B_1(x_1)$. □

It remains to consider the case that $p = 2$ and $a \in \mathbb{Z}_2^\times$. In this case, $|f'(-a)|_2 = \frac{1}{2} < 1$ for $|a|_2 = 1$. The result can be summarized as follows:

Theorem 4. *Let* $x_1 = -a$. *If* $|a|_2 = 1$, *then* x_1 *is attractive. Moreover,*

$$\mathscr{A}(x_1) = B_1(x_1).$$

Proof. Consider $|a|_2 = 1$. Condition (14.5) becomes

$$\max\{|2a|_2 r, r^2\} < 1 \Leftrightarrow \max\{\frac{r}{2}, r^2\} < 1$$

and is satisfied if $r < 1$. It follows from Corollary 1 that

$$B_r(x_1) \subset \mathscr{A}(x_1),$$

for every $r < 1$.

Suppose $|x+a|_2 \geq 1$. Then $|x|_2 \neq 1$. We have

$$\begin{aligned} |f(x)+a|_2 &= |x^3+ax^2+x+a|_2 = |(x^3+ax^2)+(x+a)|_2 \\ &= \max\{|x|_2^2|x+a|_2, |x+a|_2\} \geq |x+a|_2. \end{aligned}$$

We conclude that $\mathscr{A}(x_1) = B_1(x_1)$. □

Remark 2. Note that $|x_0 - x_1|_p = |a|_p$. So if $|a|_p < 1$, then $x_0 \in B_1(x_1) = \mathscr{S}\mathscr{I}(x_1)$. Therefore,

$$B_1(x_1) = B_1(x_0) = \mathscr{S}\mathscr{I}(x_1),$$

since every point of a ball is its center for the nonarchimedean topology. We conclude that

$$\mathscr{S}\mathscr{I}(x_1) \subset \mathscr{S}\mathscr{I}(x_0),$$

for every $a \in \mathbb{Q}_p$ such that $|a|_p < 1$ and $p \neq 2$.

References

1. Albeverio, C., Khrennikov, A., Tirozzi, B., De Smedt, S.: p-adic dynamical systems. Theor. Math. Phys. **114**, 276–287 (1998)
2. Dremov, V., Shabat, G., Vytnova, P.: On the chaotic properties of quadratic maps over non-archimedean fields. In: Khrennikov, A.Y., Rakic, Z., Volovich, I.V. (eds.) p-adic Mathematical Physics, AIP Conference Proceedings, vol. 826 pp. 43–54. Melville, New York (2006)
3. Khrennikov, A.Y., Nilsson, M.: On the number of cycles of p-adic dynamical systems. J. Number Theory **90**(2), 255–264 (2001)
4. Khrennikov, A.Y., Nilsson, M.: p-adic Deterministic and Random Dynamics. Kluwer, Dordrecht (2004)
5. Mukhamedov, F., Mendes, J.F.: On chaos of a cubic p-adic dynamical system. In: Staicu, V. (ed.) Progress in Nonlinear Differential Equations and Their Applications, AIP Conference Proceedings, vol. 75, pp. 305–316. Birkhäuser, Basel (2007)

6. Mukhamedov, F., Rozikov, U.A.: On rational p-adic dynamical systems. Methods Funct. Anal. Topol. **10**, 21–31 (2004)
7. Nilsson, M.: Cycles of monomial and perturbated monomial p-adic dynamical systems. Ann. Math. Blaise Pascal **7**(1), 37–63 (2000)
8. Shabat, B.: p-adic entropies of logistic maps. Proc. Steklov Math. Inst. **245**, 257–263 (2004)
9. Silverman, J.: The Arithmetic of Dynamical Systems. Springer, New York (2007)
10. Vladimirov, V.S., Volovich, I.V., Zelenov, E.I.: p-adic Analysis and Mathematical Physics. World Scientific, Singapore (1994)

Chapter 15
A Dynamical Approach to Congruences: Linking Circle Maps and Aperiodic Necklaces

Cristina Serpa and Jorge Buescu

Abstract There are many possible proofs of Fermat's little theorem. Among them we exemplify those using necklaces and dynamical systems. Both methods lead to a generalization. It is a congruence theorem that is already known from Gauss and Gegenbauer. A natural result from these proofs is a bijection between aperiodic necklaces and circle maps.

15.1 Introduction and Definitions

The use of dynamical systems to study and prove results in other areas of mathematics, including discrete mathematics, has been the subject of recent work. The case of the theory of numbers is a good example (see, e.g., [3,4]).

In fact, it is possible to obtain the congruence of Fermat's little theorem by counting periodic orbits and points of certain circle maps. The same method can be used to obtain an extended generalization that has been known since 1863, when a posthumous paper of Gauss was published. Its usual notation is due to Gegenbauer. This result is more general than the usual extension, namely Euler's theorem.

We can find proofs of these results using other methods, such as combinatorics. In 1872, a proof of Fermat's little theorem was given by Petersen with an argument involving counting colored boxes arranged in a circle. This idea can be seen as well as a starting point for applying sequences of numbers or letters of alphabets, that is, for obtaining the same results through working on combinatorics on words.

The key concepts we will use from combinatorics on words are those of words and necklaces (see, e.g., [1,7]).

C. Serpa (✉) • J. Buescu
Faculdade de Ciências da Universidade de Lisboa and CMAF, Lisbon, Portugal
e-mail: mcserpa@fc.ul.pt; jbuescu@ptmat.fc.ul.pt

C. Grácio et al. (eds.), *Nonlinear Maps and their Applications*, Springer Proceedings in Mathematics & Statistics 57, DOI 10.1007/978-1-4614-9161-3_15,

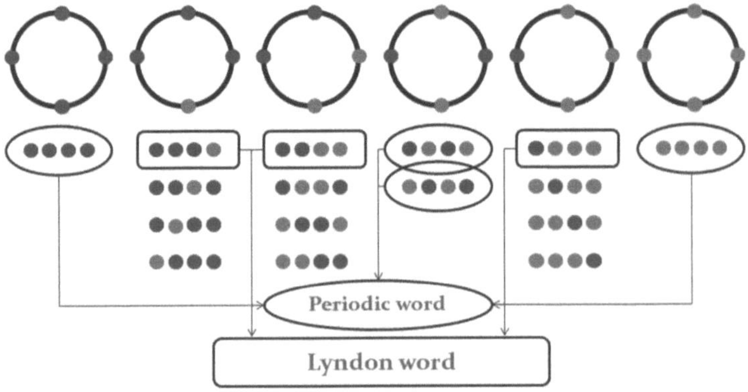

Fig. 15.1 Necklaces of length 4 from an alphabet with two symbols

Definition 1. A *word* is a finite sequence of symbols (letters) taken from a finite totally ordered set called an *alphabet*.

Definition 2. A word that can be decomposed into an n-fold repeated sequence of one word, where $n \geq 2$, is called a *periodic word*; otherwise, it is a *primitive* or *aperiodic word*.

Definition 3. The set of words for which each element can be related to every other element by a cyclic (circular) permutation of letters is an equivalence class called a *necklace*. We say that a word is the *representative word* of a necklace if it is the smallest word of the necklace in terms of the lexicographic ordering. A necklace is periodic if its representative word is periodic and is aperiodic otherwise.

Definition 4. A *Lyndon word* is a representative word of an aperiodic necklace.

Example 1. Figure 15.1 illustrates necklaces from an alphabet of two symbols.

15.2 Fermat's Little Theorem and a Generalization

Theorem 1 (Fermat's little theorem). *Let p be a prime number and let a be an integer not divisible by p. Then*

$$p \mid \left(a^{p-1} - 1\right). \tag{15.1}$$

Although this is Fermat's original formulation, the next congruence is usually referred to as Fermat's little theorem.

Theorem 2. *Let p be a prime number and let a be an integer. Then*

$$a^p \equiv a \,(mod\ p). \tag{15.2}$$

Note that unlike Theorem 1, Theorem 2 does not require that $(a,p) = 1$.

We present first a proof using necklaces.

Proof. (#1) Let p symbols form a sequence from an alphabet of a letters. If the symbols can be repeated, there are a^p possible p-sequences of symbols. A first group of possibilities is formed by the sequences that include the same symbol p times. This group has a possibilities. The other $a^p - a$ possible sequences can be grouped in sets (necklaces) of p elements, since there are p possibilities of cyclic (circular) permutations for each p-sequence of symbols. □

An analogous proof was given by Petersen in 1872 with coloring boxes (see [9]). Next we sketch a proof using dynamical systems (for more details, see, e.g., [5]).

Proof. (#2 - sketch) Let $T_a : [0,1] \to [0,1]$ defined by

$$\begin{cases} ax\,(\text{mod } 1) & \text{if } x \neq 1 \\ 1 & \text{if } x = 1. \end{cases}$$

The proof consists of the following steps:

(a) Verify that T_a has a fixed points, provided $a > 1$.
(b) Verify that $\forall a, b \in \mathbb{N}, T_a(T_b) = T_{ab}$.
(c) Count the p-periodic points of T_a, which are the fixed points of T_{a^p}.

The conclusion comes from the fact that T_{a^p} has a^p fixed points, and of these, there are exactly a fixed points of T_a. This implies that there are exactly $a^p - a$ points of T_a with minimal period p, which implies that $p | (a^p - a)$. □

Remark 1. The dynamical proof may be achieved using the circle map $g_a : [0,1) \to [0,1)$ defined by

$$g_a(x) = ax\,(\text{mod } 1), \tag{15.3}$$

requiring a slightly different way of counting the fixed points.

Both methods (combinatorial and dynamical) can be applied for any positive n (not necessarily prime) with slightly more refined methods. The following generalization of Fermat's little theorem is a consequence of these results.

Theorem 3. *Let $a, n \in \mathbb{N}$. Then*

$$\sum_{d|n} \mu(d)\, a^{n/d} \equiv 0\ (mod\ n). \tag{15.4}$$

Here μ is the Möbius function, which for $n = p_1^{\alpha_1} p_2^{\alpha_2} \cdots p_k^{\alpha_k}$ is defined as follows:

$$\mu(n) = \begin{cases} 1 & \text{if } n = 1, \\ 0 & \text{if there exists } j \in \{1, \ldots, k\} \text{ such that } \alpha_j \geq 2 \\ (-1)^k & \text{if } \alpha_1 = \alpha_2 = \cdots = \alpha_k = 1. \end{cases} \tag{15.5}$$

This result is due to Gauss, and the notation is from Gegenbauer (see [2]).

Example 2. For $n = 60$ and $a \in \mathbb{N}$,

$$a^{60} - a^{30} - a^{20} - a^{12} + a^{10} + a^{6} + a^{4} - a^{2} \equiv 0 \ (\bmod 60). \tag{15.6}$$

15.3 Linking Circle Maps and Aperiodic Necklaces

A common property of the above-mentioned proofs of the generalization of Fermat's little theorem is the fact that there are sets of the same cardinality that apparently have nothing in common.

The next result links the theories of combinatorics and dynamical systems.

Theorem 4. *Let $a, n \in \mathbb{N}$. The number of necklaces of length n in an alphabet of cardinality a and the number of all periodic orbits of minimum period of the circle map g_a is*

$$\frac{1}{n} \sum_{d|n} \mu(d)\, a^{n/d}. \tag{15.7}$$

Consequently, we can construct a bijection between the set of all necklaces of length n out of an alphabet of cardinality a and the set of all n-periodic orbits of the circle map g_a. Here n is the minimum period of the orbit.

For this purpose, we introduce some definitions, adapting the notation of Milnor and Thurston [6].

Definition 5. Let $a \in \mathbb{N}$. The circle map g_a defined from $[0,1)$ to itself is called *piecewise monotone* because $[0,1)$ can be subdivided into finitely many intervals $J_1, \ldots, J_a$ on which g_a is continuous and strictly increasing or strictly decreasing. Each such maximal interval on which g_a is monotone is called a *lap* of f. Here $J = [c_0, c_a]$, $J_j = [c_j, c_{j+1}]$ $\forall j \in \{0, 1, \ldots, a-1\}$, and $c_0 < c_1 < \cdots < c_a$. The separating interior points $c_1, \ldots, c_{a-1}$ at which f is a local minimum or maximum are called the *turning points* of f.

Definition 6. The *address* $A(x)$ of a point $x \in [0,1)$ is the formal symbol J_j if x belongs to the lap J_j and is not a turning point, or the formal symbol c_j if x is precisely equal to the turning point c_j. The *itinerary* $I(x)$ is the sequence of addresses $\left(A(x), A(g_a(x)), A\left(g_a^2(x)\right), \ldots\right)$ of the successive images of x. By $I_n(x)$ we denote the n-length string corresponding to truncating $I(x)$ at the symbol n, and call it the *n-truncated itinerary*.

Theorem 5. *Let $a, n \in \mathbb{N}$. Let $x \in [0,1)$ be a periodic point of minimum period n of g_a. Then the n-truncated itinerary of x, considered as a word from the alphabet of symbols $J_1, \ldots, J_a$, is an aperiodic word.*

The proof is given in [8].

Therefore, the bijection is constructed by the identification of a periodic point with its n-truncated itinerary, where n is its minimum period. As a consequence, we

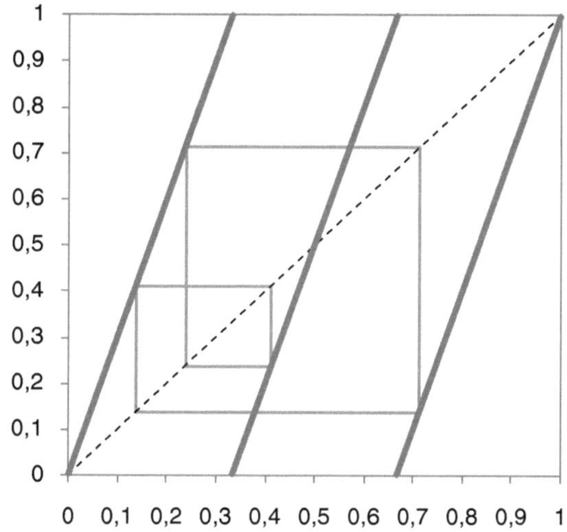

Fig. 15.2 An orbit of g_3

can identify each periodic orbit with an aperiodic necklace, which is the same as identifying it with a Lyndon word.

Example 3. If we use the symbols 0, 1, and 2 respectively to denote the subintervals $\left[0, \frac{1}{3}\right)$, $\left[\frac{1}{3}, \frac{2}{3}\right)$, and $\left[\frac{2}{3}, 1\right)$. Figure 15.2 represents the Lyndon word 0102.

In fact, it is possible to obtain all the aperiodic words of length n over a finite alphabet of cardinality a through all the n-periodic orbits of the corresponding circle map. Conversely, having an aperiodic word (in particular a Lyndon word), we can compute the points of a periodic orbit of the circle map. More details are in an article to be published by the authors [8].

Acknowledgements JB acknowledges partial support from FCT through Financiamento Base 2010-ISFL-1-209. CS acknowledges support from FCT through grant SFRH/77623/2011.

References

1. Berstel, J., Perrin, D.: The origins of combinatorics on words. Euro. J. Comb. **28**, 996–1022 (2007)
2. Dickson, L.E.: History of the Theory of Numbers: Divisibility and Primality, vol. 1. Carnegie Institution of Washington (publication no 256), Washington (1919)
3. Furstenberg, H.: Recurrence in Ergodic Theory and Combinatorial Number Theory. Princeton University Press, Princeton (1981)
4. Furstenberg, H., Katznelson, Y., Ornstein D.: The ergodic theoretical proof of Szemerédi's theorem. Bull. Amer. Math. Soc. **7**(3), 527–552 (1982)
5. Iga, K.: A Dynamical Systems proof of Fermat's little theorem. Math. Mag. **76**(1), 48–51 (2003)

6. Milnor, J., Thurston, W.: On Iterated Maps of the Interval. In: Lectures Notes on Mathematics, vol. 1342, pp. 465–563. Springer, Berlin (1988)
7. Rozenberg, G., Salomaa, A.: Handbook of Formal Languages: Word Language Grammar, vol. 1. Springer, New York (1997)
8. Serpa, C., Buescu, J.: Combinatorics of piecewise expanding interval maps (submitted)
9. Smyth, C.J.: A coloring proof of a generalization of Fermat's little theorem. The Amer. Math. Month. **93**(6), 469–471 (1986)

Chapter 16
Complex Patterns in a Chain of Coupled Maps Based on a Neuron Model with Space and Time-Varying Couplings

Yoko Uwate and Yoshifumi Nishio

Abstract Biological neurons are able to exhibit spiking and bursting behavior. The interesting phenomena can be observed by synchronization of neuronal ensembles. We study synchronization phenomena in a chain of 2-dimensional maps based on a neuronal model with space-varying and time-varying couplings. In the case of space-varying coupling, wave propagation and complex patterns can be observed.

16.1 Introduction

Generally, complex dynamical phenomena can be observed in networks formed by many elements with nonlinearity. A coupled map lattice (CML) has been proposed by Kaneko and Bunimovich [1–5] to be used as general models for the complex high-dimensional dynamics, such as biological systems, networks in DNA, economic activities, neural networks, and evolution. We can observe spatiotemporal patterns in CML. Moreover, coupled oscillatory systems can also produce interesting phase patterns, including wave propagation, clustering, and complex phase patterns [6–9]. It is very important to clarify the mechanism of these spatiotemporal patterns if we are to understand the complex patterns observed in nature. Usually, chaotic maps are used for CML, and many types of spatiotemporal patterns have been observed.

Recently, a discrete map for spiking–bursting neural behavior was proposed by Rulkov (Fig. 16.1) [10, 11]. The Rulkov map in the form of a two-dimensional map can be useful for understanding the dynamical mechanism of oscillators in large-scale networks. And the Rulkov map produces the spiking–bursting behavior observed in real neurons. In this study, we use Rulkov maps applied to CML to study synchronization in large-scale networks. Furthermore, we assume that the coupling

Y. Uwate (✉) • Y. Nishio
Tokushima University, 2-1 Minami Josanjima, Tokushima 770-8506, Japan
e-mail: uwate@ee.tokushima-u.ac.jp; nishio@ee.tokushima-u.ac.jp

C. Grácio et al. (eds.), *Nonlinear Maps and their Applications*, Springer Proceedings in Mathematics & Statistics 57, DOI 10.1007/978-1-4614-9161-3_16,

strength between the neurons is not simple and that the coupling strength plays an important role for the whole system. We consider that the coupling strength changes with respect to space and time. Space-varying coupling is realized using a normal distribution. And we realize time-varying coupling by periodically changing values between positive and negative.

In this study, we investigate synchronization phenomena in a chain of coupled Rulkov maps with space-varying and time-varying couplings. When space-varying coupling is used, irregular wave propagation can be observed, while in the case of time-varying coupling, complex patterns are observed. We confirm that more complex patterns are generated by changing the coupling value with time.

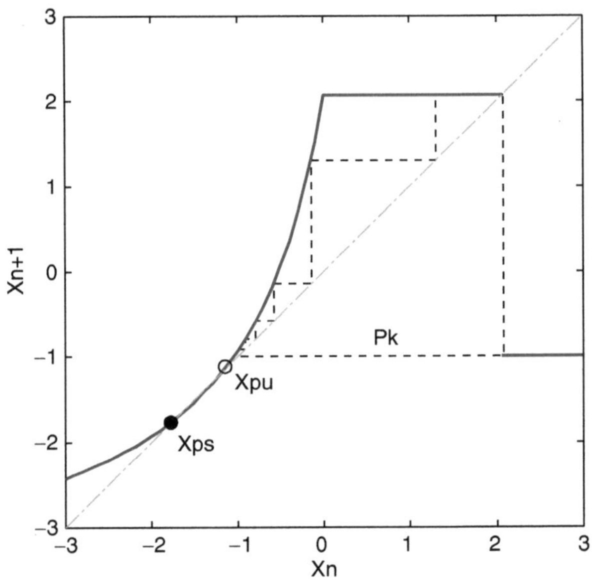

Fig. 16.1 Rulkov map. The *dashed line* illustrates a superstable cycle P_k. The stable and unstable fixed points of the map are indicated by x_{ps} and x_{pu}, respectively

16.2 Two Coupled Rulkov Maps

Consider two coupled Rulkov maps [10] as given by the following equation:

$$\begin{aligned} x_{i,n+1} &= f(x_{i,n}, y_{i,n} + \beta_{i,n}), \\ y_{i,n+1} &= y_{i,n} - \mu(x_{i,n} + 1) + \mu\sigma_i + \mu\sigma_{i,n}, \end{aligned} \tag{16.1}$$

$$f(x_n, y) = \begin{cases} \alpha/(1 - x_n) + y, & x_n \leqslant 0, \\ \alpha + y, & 0 < x_n < \alpha + y \text{ and } x_n \leqslant 0, \\ -1, & x_n \geqslant \alpha + y \text{ or } x_{n-1} > 0, \end{cases} \tag{16.2}$$

where x and y are the fast and slow dynamical variables, respectively. The coupling between the cells is provided by the current flowing from one cell to another. This coupling is modeled by

$$\begin{aligned} \beta_{i,n} &= g\beta^e(x_{j,n} - x_{i,n}), \\ \sigma_{i,n} &= g\sigma^e(x_{j,n} - x_{i,n}), \end{aligned} \tag{16.3}$$

where g denotes the coupling strength. In numerical simulations, the values of the coefficients are set equal: $\beta^e = 1.0$ and $\sigma^e = 1.0$. The other parameters have the following values: $\mu = 0.001$, $\alpha = 5.0$, and $\sigma = 0.24$. The coupling between the maps is symmetric, i.e., $g_{ji} = g_{ij} = g$.

First, we investigate basic synchronization phenomena when the normal coupling is used without changing the value of the coupling. When two Rulkov maps are coupled with positive value $g = 0.029$, two bursting waves are synchronized at the in-phase, as shown in Fig. 16.2a, while introduction of negative coupling $g = -0.029$ in this regime of synchronization shows an antiphase (see. Fig. 16.2b). For comparison between the in-phase and the antiphase states, the oscillation frequency of the antiphase is faster than that of the in-phase state.

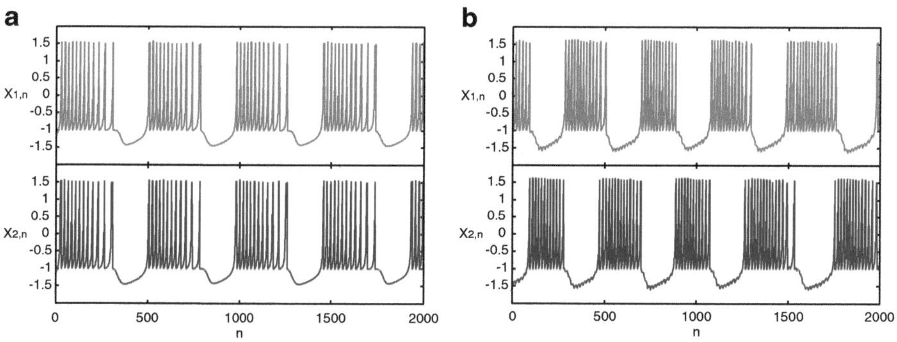

Fig. 16.2 In-phase/antiphase synchronization. (**a**) In-phase ($g = 0.029$). (**b**) Antiphase ($g = -0.029$)

When three maps are coupled with the negative coupling, we observe the three-phase synchronization as shown in Fig. 16.3.

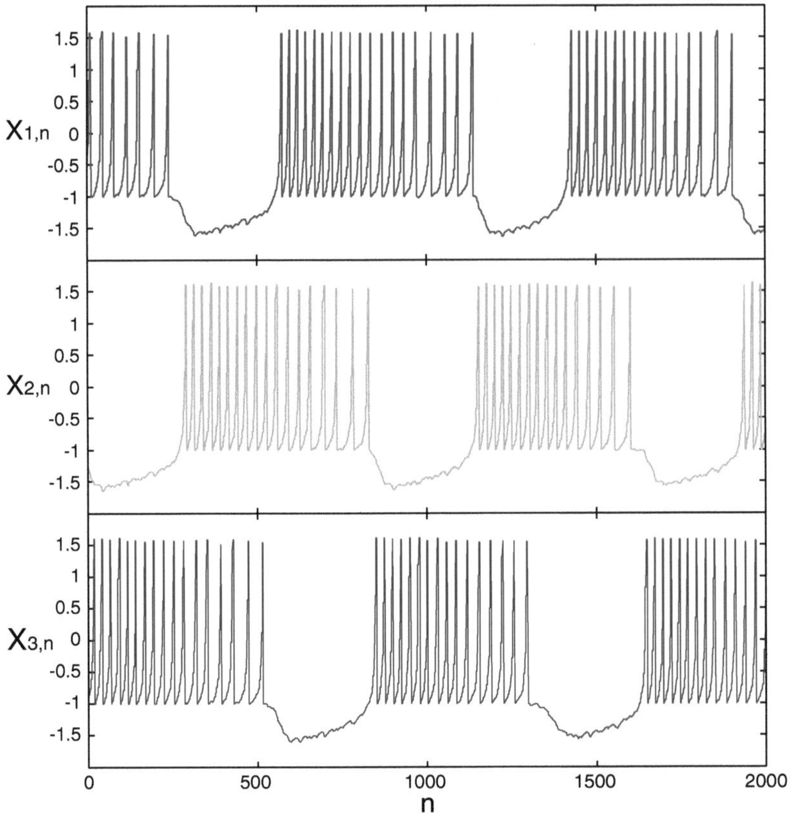

Fig. 16.3 Three-phase synchronization ($g = -0.029$)

16.3 Synchronization in a Chain of Coupled Maps

We consider a chain of coupled maps (see Fig. 16.4):

$$\begin{aligned} x_{i,n+1} &= f(x_{i,n}x_{i,n-1}, y_{i,n}) \\ &\quad + \frac{1}{2}g(x_{i+1,n} - 2x_{i,n} + x_{i-1,n}), \\ y_{i,n+1} &= y_{i,n} - \mu(x_{i,n} + 1) + \mu\sigma_i \\ &\quad + \mu\frac{1}{2}g(x_{i+1,n} - 2x_{i,n} + x_{i-1,n}), \\ &\quad i = 1, \ldots, N, \end{aligned} \tag{16.4}$$

where x and y are the fast and slow dynamical variables, respectively, $\mu = 10^{-3}$ and σ_i are the parameters of the individual map, and g is the coupling. The function $f()$ has the following form:

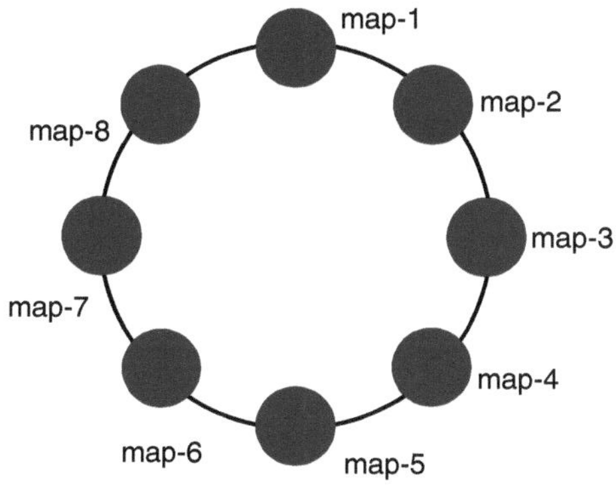

Fig. 16.4 Conceptual chain of coupled maps ($N = 8$)

$$f(x_n, y_n) = \begin{cases} \alpha/(1 - x_n) + y_n, & x_n \leq 0, \\ \alpha + y_n, & 0 < x_n < \alpha + y_n \text{ and } x_{n-1} \leq 0, \\ -1, & x_n \geq \alpha + y_n \text{ or } x_{n-1} > 0, \end{cases} \tag{16.5}$$

In this simulation, we take $\alpha = 3.5$, and σ_i is randomly distributed in the interval $[0.15, 0.16]$. The number of coupled maps is set to 100.

16.3.1 Normal Coupling

For comparison, we show synchronization of a chain of Rulkov maps when the coupling strength is constant in space and time. The simulation results of the space–time plot of positive and negative constant coupling are shown in Figs. 16.5a and b, respectively. The horizontal axis is iteration time n, and the vertical axis is space i. The wave propagation can be observed when the space-varying coupling has positive value. In the case of negative coupling, a complex pattern is generated.

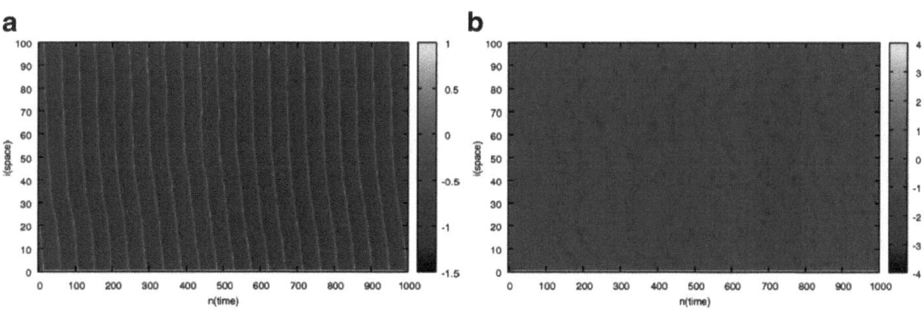

Fig. 16.5 Space-time plots of x_i. (**a**) Positive constant coupling ($g = 0.38$) (**b**) Negative constant coupling ($g = -0.38$)

16.3.2 Space-Varying Coupling

In this study, space-varying coupling is realized using a normal distribution. The normal distribution is given by the formula

$$f(z) = \frac{1}{\sqrt{2\pi}\sigma} \exp\left(-\frac{(z-\mu)^2}{2\sigma^2}\right), \tag{16.6}$$

where μ is the mean and σ is the standard deviation. An example of space-varying coupling is shown in Fig. 16.6.

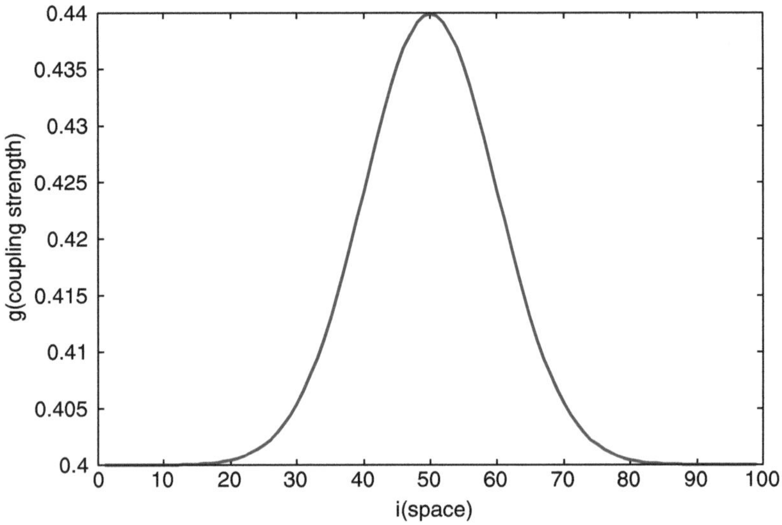

Fig. 16.6 Space-varying coupling ($\mu = 50$, $\sigma = 10$)

First, we consider that the space-varying coupling has positive value. The simulation results by changing the initial condition of the maps are shown in Fig. 16.7. From these figures, we can see that irregular wave propagation can occur. The form of irregular wave propagation depends on the initial condition of the maps.

Next, we investigate the synchronization of maps when the space-varying coupling has negative value. In this case, we change σ of the normal distribution to study the influence of the forms of space-varying coupling. Fig. 16.8 shows space-varying coupling with negative value resulting from changing σ. The space–time plots are shown in Fig. 16.9. We confirmed the high density of waves when coupling is of high amplitude.

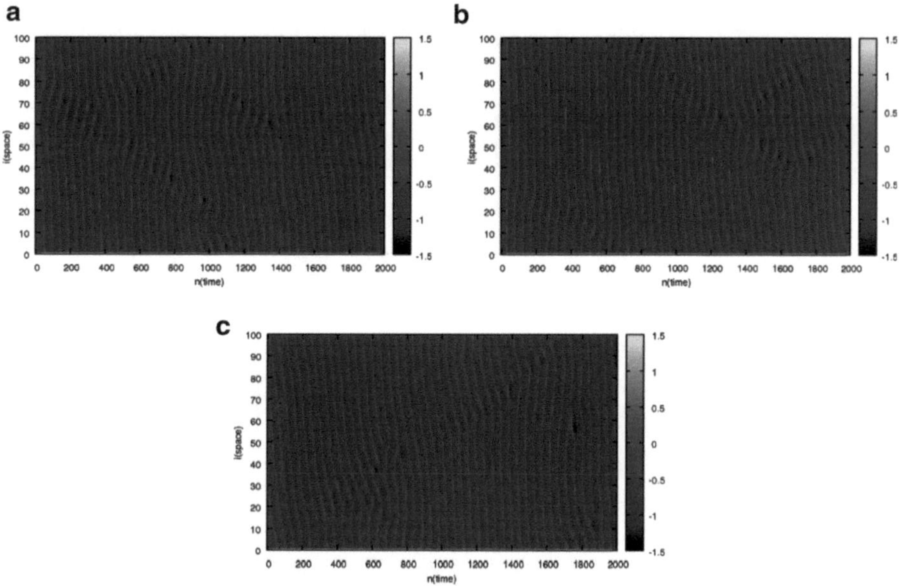

Fig. 16.7 Examples of regular wave propagation with irregular wave propagation ($\mu = 50$, $\sigma = 10$) (a) Example 1, (b) Example 2, (c) Example 3

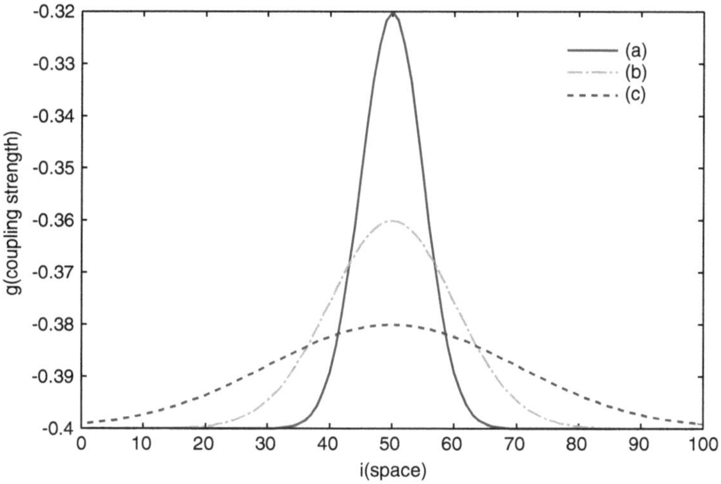

Fig. 16.8 Space-varying coupling with negative value by changing σ. (a) $\sigma = 5$, (b) $\sigma = 10$, (c) $\sigma = 20$

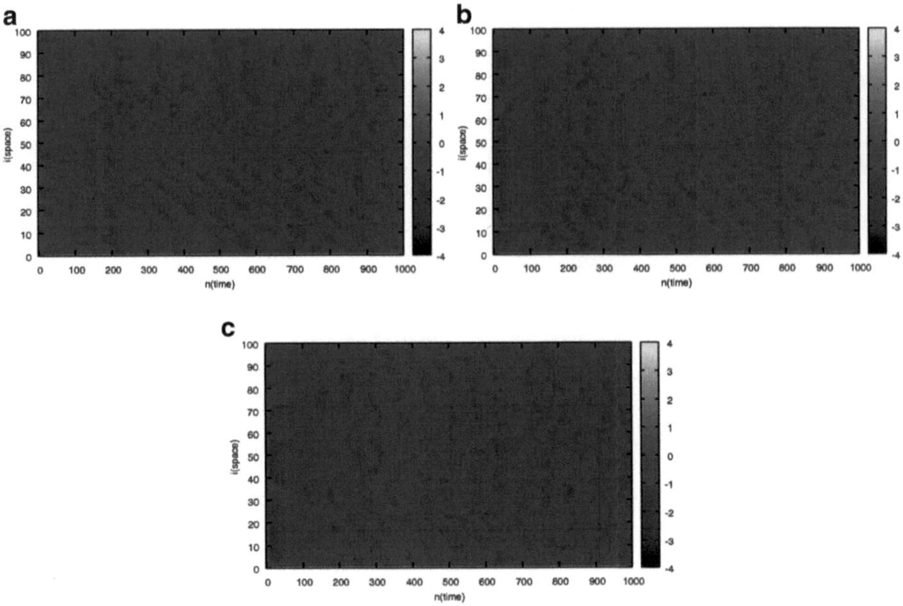

Fig. 16.9 Complex patterns with space-varying coupling (negative value) (**a**) $\sigma = 5$, (**b**) $\sigma = 10$, (**c**) $\sigma = 20$

16.3.3 Time-Varying Coupling

Next, we consider time-varying coupling when two maps are coupled. Figure 16.10 shows the characteristics of the time-varying coupling. The value of the time-varying coupling switches periodically between positive and negative, and p denotes the switching period.

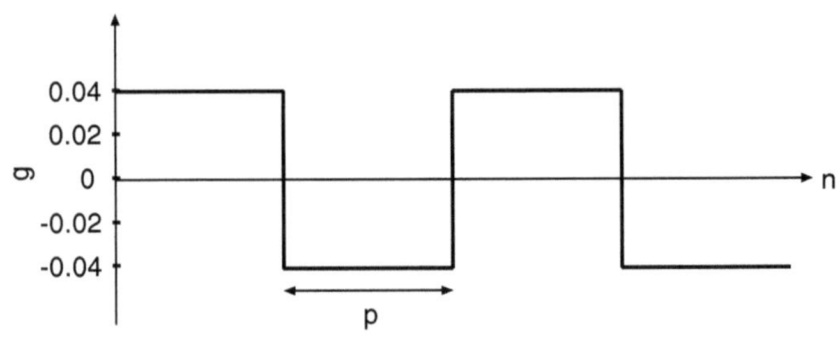

Fig. 16.10 Characteristics of time-varying coupling

Using time-varying coupling, we observe the synchronization in switching between in-phase and anti-phase depending on the coupling strength (when the switching period is set to $p = 1000$, $p = 200$) as shown in Figs. 16.11 and 16.12.

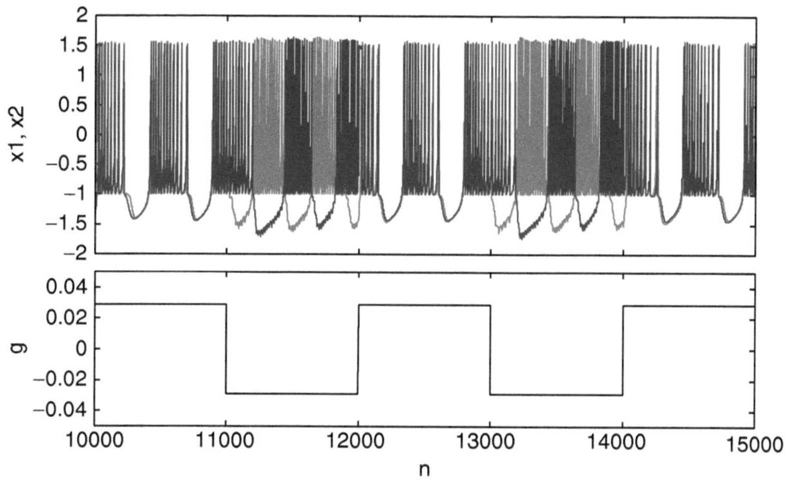

Fig. 16.11 Switching of in-phase and anti-phase synchronization ($p = 1000$)

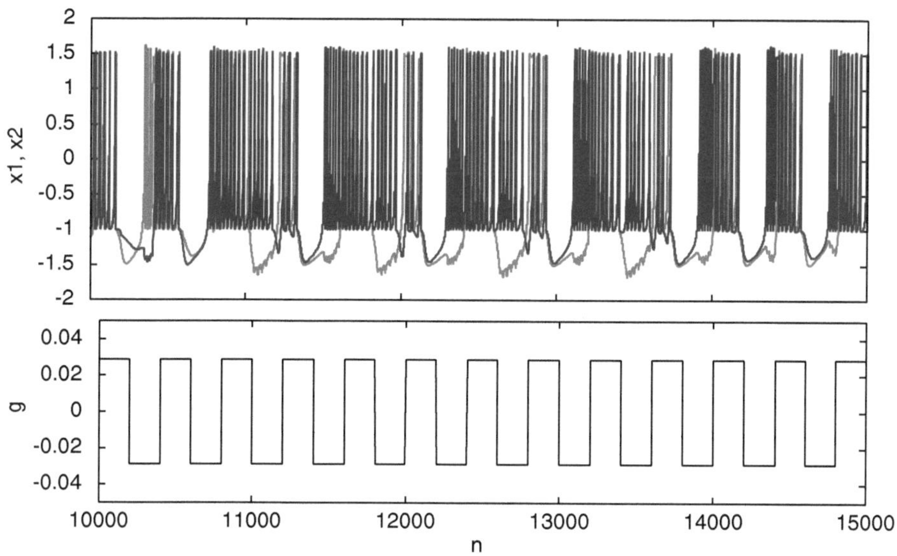

Fig. 16.12 Switching of in-phase and anti-phase synchronization ($p = 200$)

The upper figure shows the two waveforms generated from two coupled Rulkov maps, and the lower figure shows the characteristics of the time-varying coupling. When the coupling strength has positive value, the two maps are synchronized in in-phase, while when the coupling value is negative, the two maps are synchronized in antiphase. By switching the value of the time-varying coupling, the synchronization state also switches between the in-phase and the antiphase states.

Next, we calculated the frequency of the bursting part when the iteration time is set to $n = 1\,000\,000$, as shown in Fig. 16.13. From this result we can confirm that in the case of normal positive coupling, only three peaks exist. In contrast, there are many peaks when negative coupling is used. When we use time-varying coupling, several burst peaks are observed.

Figure 16.14 shows the simulation result of the average burst length by changing the period of the time-varying coupling. We confirmed two peaks of the period of time-varying coupling in this figure. We confirm that the average burst length decreases when the period of time-varying coupling becomes large.

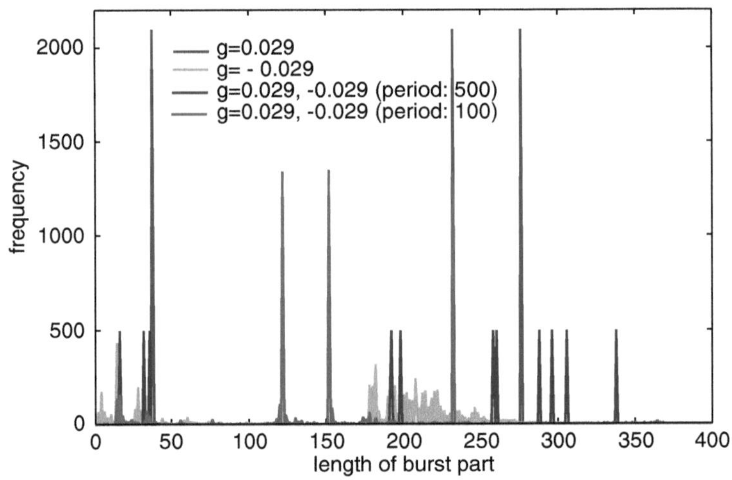

Fig. 16.13 Histogram of the burst length

Next, we investigate synchronization in the chain of coupled maps. The chain's model is the same as the model presented in Sect. 16.3 The simulation results of the space–time plot by changing the period of time-varying coupling are shown in Fig. 16.15a–c.

When the time-varying coupling is set to positive, wave propagation can be observed, while in the case of negative coupling, we confirm that random patterns are produced. By decreasing the period of time-varying coupling, the length of each region of wave propagation and random patterns becomes shorter. Figure 16.15d shows the simulation result of the space–time plot when the coupling is set to $g = \pm 0.4$ and the period of time-varying coupling is $p = 100$.

We see that interesting complex patterns can be observed by changing the coupling strength.

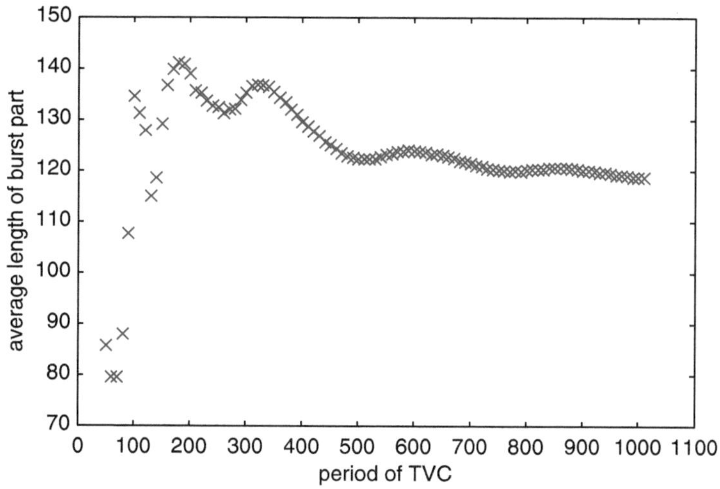

Fig. 16.14 Average burst length

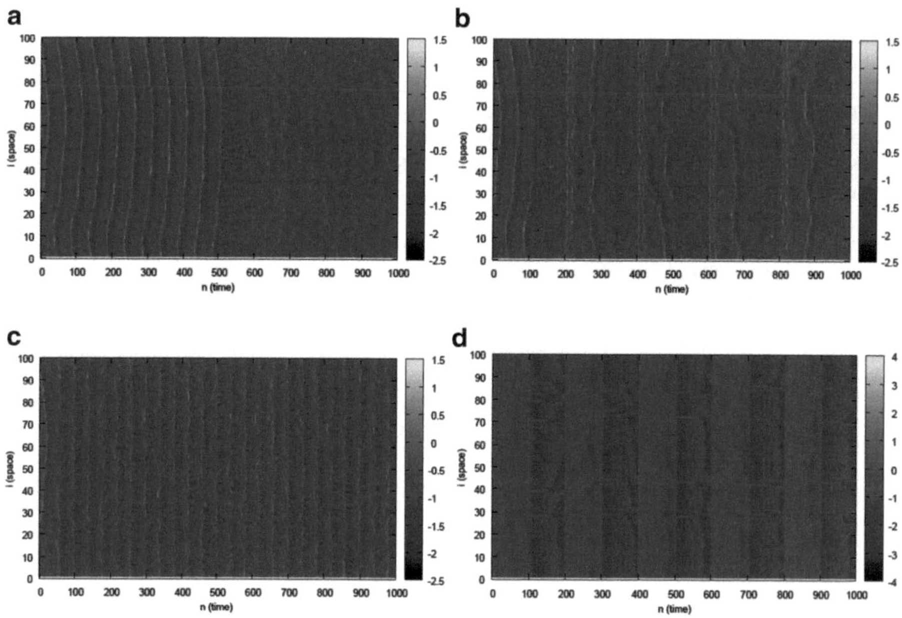

Fig. 16.15 Space–time plots obtained by changing the parameters p and g. (**a**) x_i $(g = \pm 0.2, p = 500)$, (**b**) x_i $(g = \pm 0.2, p = 100)$, (**c**) x_i $(g = \pm 0.2, p = 20)$, (**d**) x_i $(g = \pm 0.4, p = 100)$

16.4 Conclusions

In this paper, we have studied synchronization in a chain of coupled Rulkov maps with space-varying and time-varying couplings. We have realized space-varying coupling using a normal distribution. When space-varying coupling is used, irregular wave propagation was observed, while in the case of time-varying coupling, complex patterns were observed.

In the future works, we shall study these complex phenomena by applying statistical analysis, and we shall investigate the synchronization of maps by other coupling methods such as chemical and diffusion couplings.

References

1. Kaneko, K.: Spatiotemporal intermittency in coupled map lattice. Prog. Theor. Phys. **75**(5), 1033–1044 (1985)
2. Bunimovich, L.A., Sinai, Ya. G.: Spacetime chaos in coupled map lattices. Nonlineality **1**(4), 491–516 (1988)
3. Kaneko, K.: Pattern dynamics in spatiotemporal chaos. Physica D **34**, 1–41 (1989)
4. Kaneko, K.: Spatiotemporal chaos in one- and two-dimensional coupled map lattice. Physica D **37**, 60–82 (1989)
5. Kaneko, K.: Simulating physics with Coupled Map Lattice–Pattern Dynamics, Information Flow, and Thermodynamics of Spatiotemporal Chaos, Formation, Dynamics, and Statistics of Patterns, pp. 1–52. World Scientific (1990)
6. Nishio, Y. Mori, S.: Mutually Coupled Oscillators with an Extremely Large Number of Steady States, vol. 2, pp. 819–822. Proceedings of ISCAS'92 May (1992)
7. Pivka, L.: Autowaves and spatiotemporal chaos in CNNs Part I: A tutorial. IEEE Trans. Circuits Syst. I **42**, 638–649 (1995)
8. Yamauchi, M., Nishio, Y., Ushida, A.: Phase-waves in a ladder of oscillators IEICE Trans. Fundamentals **E86-A**(4), 891–899 (2003)
9. Uwate, Y., Nishio, Y.: Complex Phase Synchronization in an Array of Oscillators Coupled by Time-Varying Resistor, pp. 8345–8350. Porc. of IJCNN'06, July (2006)
10. Rulkov, N.F.: Modeling of spiking–bursting neural behavior using two-dimensional map. Physical Rev., E **65**, 041922 (2002)
11. Rulkov, N.F., Timofeev, I., Bazhenov, M.: Oscillations in large-scale cortical networks: map-based model. J. Computat. Neurosci. **17**, 203–223 (2004)

MIX
Papier aus verantwortungsvollen Quellen
Paper from responsible sources
FSC® C105338

If you have any concerns about our products,
you can contact us on
ProductSafety@springernature.com

In case Publisher is established outside the EU,
the EU authorized representative is:
Springer Nature Customer Service Center GmbH
Europaplatz 3, 69115 Heidelberg, Germany

Printed by Libri Plureos GmbH
in Hamburg, Germany